BIORESOURCES TECHNOLOGY IN SUSTAINABLE AGRICULTURE

Biological and Biochemical Research

BIORESOURCES TECHNOLOGY IN SUSTAINABLE AGRICULTURE

Biological and Biochemical Research

Edited by

Mohamad Faiz Foong Abdullah, DPhil
Mohd Tajudin Bin Mohd Ali, PhD
Farida Zuraina M. Yusof, PhD

AAP | APPLE ACADEMIC PRESS

Apple Academic Press Inc.	Apple Academic Press Inc.
3333 Mistwell Crescent	9 Spinnaker Way
Oakville, ON L6L 0A2 Canada	Waretown, NJ 08758 USA

© 2018 by Apple Academic Press, Inc.

First issued in paperback 2021

No claim to original U.S. Government works

ISBN 13: 978-1-77-463042-6 (pbk)
ISBN 13: 978-1-77-188449-5 (hbk)

Library and Archives Canada Cataloguing in Publication

Bioresources technology in sustainable agriculture: biological and biochemical research / edited by Mohamad Faiz Foong Abdullah, DPhil, Mohd Tajudin Bin Mohd Ali, PhD, Farida Zuraina M. Yusof, PhD.
Includes bibliographical references and index.
Issued in print and electronic formats.
ISBN 978-1-77188-449-5 (hardcover).--ISBN 978-1-315-36596-1 (PDF)
1. Agricultural biotechnology. 2. Agricultural chemistry. 3. Sustainable agriculture--Technological innovations. 4. Agricultural innovations. I. Abdullah, Mohamad Faiz Foong, editor
S494.5.B563B52 2017   630   C2017-906507-6   C2017-906508-4

Library of Congress Cataloging-in-Publication Data

Names: Abdullah, Mohamad Faiz Foong, editor. | Ali, Mohd Tajudin Bin Mohd, editor. | Yusof, Farida Zuraina M., editor.
Title: Bioresources technology in sustainable agriculture : biological and biochemical research / editors, Mohamad Faiz Foong Abdullah, DPhil, Mohd Tajudin Bin Mohd Ali, PhD, Farida Zuraina M. Yusof, PhD.
Description: Waretown, NJ: Apple Academic Press, 2017. | Includes bibliographical references and index.
Identifiers: LCCN 2017045042 (print) | LCCN 2017045998 (ebook) | ISBN 9781315365961 (ebook)
| ISBN 9781771884495 (hardcover : alk. paper)
Subjects: LCSH: Agricultural biotechnology--Research. | Agricultural resources--Research.
Classification: LCC S494.5.B563 (ebook) | LCC S494.5.B563 B5275 2017 (print) | DDC 630.72--dc23
LC record available at https://lccn.loc.gov/2017045042

ABOUT THE EDITORS

Mohamad Faiz Foong Abdullah, DPhil, is currently Associate Professor in the Faculty of Applied Sciences at the Universiti Teknologi MARA in Malaysia. He graduated with a degree in microbiology from the University of Malaya and completed his DPhil in molecular genetics as an Islamic Development Bank Merit Scholar at the University of Oxford, United Kingdom. Subsequently, he was awarded an international grant from the Wellcome Trust as a visiting research fellow in the University of Leicester, United Kingdom. His responsibilities include teaching, research, and administration. His main area of research is in genetic engineering and cellular biology of the model organism *Saccharomyces cerevisiae* and microbiology. He is also a member of the Genetic Modification Advisory Committee under the National Biosafety Board.

Mohd Tajudin Bin Mohd Ali, PhD, is a senior lecturer (organic chemistry) at the Universiti Teknologi MARA in Malaysia. He obtained his Bachelor of Science (Hons.) degree in Industrial Chemistry from the Universiti Putra Malaysia (UPM), followed by a Diploma in Education from Maktab Perguruan Raja Melewar; a Master in Science (natural product) from UPM; and a PhD (organic synthesis) from the University of Regensburg, Germany. He was the author of the books *Simplified Approach to the Mole Concept and Modern Methods in Organic Chemistry* and *Chemical Waste Management,* published by UPENA. He is the author of *Perancangan Sintesis Kimia Organik,* published by Dewan Bahasa dan Pustaka. He also was the author and co-author for more than 20 research papers in the areas of organic synthesis, peptide synthesis, and natural product synthesis.

Farida Zuraina M. Yusof, PhD, is currently Associate Professor in the Faculty of Applied Sciences, at the Universiti Teknologi MARA in Malaysia. She received her BSc degree in biochemistry from the Universiti Kebangsaan Malaysia, where she studied cancer research; her MPhil degree in molecular biology from the Universiti Malaya; and her PhD in medical biotechnology from Ulster University of Northern Ireland. She has authored and reviewed numerous articles on biology, biochemistry, genetics, and molecular biology. Her research work has been supported by the Ministry of Education and Ministry of Science and Innovation of Malaysia. She has taught biochemistry and molecular biology at the university for 18 years and has advised undergraduates and also postgraduates working on biochemical and molecular biology research projects. She is a member of the Malaysian Society for Applied Biology and Malaysian Society for Molecular Biology and Biochemistry.

CONTENTS

LIST OF CONTRIBUTORS

Azzura Abdullah
Faculty of Applied Sciences, Universiti Teknologi MARA, 40450 Shah Alam, Malaysia

Mohd Faiz Foong Abdullah
Faculty of Applied Sciences, Universiti Teknologi MARA, 40450 Shah Alam, Malaysia

Noor Ezan Abdullah
Faculty of Electrical Engineering, Universiti Teknologi MARA, 40450 Shah Alam, Malaysia

Mohd Rozi Ahmad
Faculty of Applied Sciences, Universiti Teknologi MARA, 40450 Shah Alam, Malaysia

Rohana Ahmad
Centre for Restorative Dentistry Studies, Faculty of Dentistry, Universiti Teknologi MARA, 40450 Shah Alam, Malaysia

Wan Yunus Wan Ahmad
Faculty of Applied Sciences, Universiti Teknologi MARA, 40450 Shah Alam, Malaysia

Mohd Tajudin bin Mohd Ali
Faculty of Applied Sciences, Universiti Teknologi MARA, 40450 Shah Alam, Malaysia

Shamsul Bahrin Gulam Ali
Faculty of Applied Sciences, Universiti Teknologi MARA, 40450 Shah Alam, Malaysia

Siti Aisyah binti Aliasak
Faculty of Applied Sciences, Universiti Teknologi MARA, 40450 Shah Alam, Malaysia

Jessey Angat
Faculty of Applied Sciences, Universiti Teknologi MARA, 40450 Shah Alam, Malaysia

Mohd Hisyam Mohd Ariff
Faculty of Electrical Engineering, Universiti Malaysia Pahang, 26600 Pekan, Pahang, Malaysia

Amizon Azizan
Faculty of Chemical Engineering, Universiti Teknologi MARA, 40450 Shah Alam, Malaysia

Azrie Faris Mohd Azmi
Faculty of Electrical Engineering, Universiti Teknologi MARA, 40450 Shah Alam, Malaysia

Tay Chia Chay
Faculty of Applied Sciences, Universiti Teknologi MARA (Perlis), 02600 Arau, Perlis, Malaysia

Wan Nurhayati Wan Hanafi
Faculty of Applied Sciences, Universiti Teknologi MARA, 40450 Shah Alam, Malaysia

Hadzli Hashim
Faculty of Electrical Engineering, Universiti Teknologi MARA, 40450 Shah Alam, Malaysia

Wan Rozianoor Mohd Hassan
Faculty of Applied Sciences, Universiti Teknologi MARA, 40450 Shah Alam, Malaysia

Wan Nazihah Wan Ibrahim
Faculty of Applied Sciences, Universiti Teknologi MARA, 40450 Shah Alam, Malaysia

Asmida Ismail
Faculty of Applied Sciences, Universiti Teknologi MARA, 40450 Shah Alam, Malaysia

Faridatul Aima Ismail
Faculty of Electrical Engineering, Universiti Teknologi MARA, 40450 Shah Alam, Malaysia

Hasnun Nita binti Ismail
Faculty of Applied Sciences, Universiti Teknologi MARA, 35400 Tapah Road, Perak, Malaysia

Ismarani Ismail
Faculty of Electrical Engineering, Universiti Teknologi MARA, 40450 Shah Alam, Selangor, Malaysia

Roslinda binti Ismail
Faculty of Applied Sciences, Universiti Teknologi MARA, 40450 Shah Alam, Malaysia

Siti Nurhazlin Jaluddin
Faculty of Applied Sciences, Universiti Teknologi MARA, 40450 Shah Alam, Malaysia

Nursuriati Jamil
Faculty of Computer and Mathematical Sciences, Universiti Teknologi MARA, 40450 Shah Alam, Malaysia

Muhammad Ismail Abd Kadir
Faculty of Applied Sciences, Universiti Teknologi MARA, 40450 Shah Alam, Malaysia

Syed Abdul Illah Alyahya Syed Abdul Kadir
Faculty of Applied Sciences, Universiti Teknologi MARA, 40450 Shah Alam, Malaysia

Karimah Kassim
Faculty of Applied Sciences, Universiti Teknologi MARA, 40450 Shah Alam, Malaysia

Nurul Fatihah Khairuddin
Faculty of Applied Sciences, Universiti Teknologi MARA, 40450 Shah Alam, Malaysia

Noor Aishah Khairuzzaman
Faculty of Electrical Engineering, Universiti Teknologi MARA, 40450 Shah Alam, Malaysia

Khalilah Abdul Khalil
Faculty of Applied Sciences, Universiti Teknologi MARA, 40450 Shah Alam, Malaysia

Tong Wah Lim
Centre for Restorative Dentistry Studies, Faculty of Dentistry, Universiti Teknologi MARA, 40450 Shah Alam, Malaysia

Nina Korlina Madzhi
Faculty of Electrical Engineering, Universiti Teknologi MARA, 40450 Shah Alam, Malaysia

Khairulmazidah binti Mohamed
Faculty of Applied Sciences, Universiti Teknologi MARA, 40450 Shah Alam, Malaysia

Tengku Elida Tengku Zainal Mulok
Faculty of Applied Sciences, Universiti Teknologi MARA, 40450 Shah Alam, Malaysia

Kamsani Ngalib
Faculty of Applied Sciences, Universiti Teknologi MARA, 40450 Shah Alam, Malaysia

Rohana Mat Nor
Faculty of Applied Sciences, Universiti Teknologi MARA, 40450 Shah Alam, Malaysia

Sharifalillah Nordin
Faculty of Computer and Mathematical Sciences, Universiti Teknologi MARA, 40450 Shah Alam, Malaysia

Muhamad Faridz Osman
Faculty of Applied Sciences, Universiti Teknologi MARA, 40450 Shah Alam, Malaysia

Wan Siti Atikah Wan Osman
Faculty of Applied Sciences, Universiti Teknologi MARA, 26400 Bandar Jengka, Malaysia

Asysyuura Adytia Patar
Faculty of Applied Sciences, Universiti Teknologi MARA, 40450 Shah Alam, Malaysia

Nik Roslan Nik Abdul Rashid
Faculty of Applied Sciences, Universiti Teknologi MARA, 40450 Shah Alam, Malaysia

Rumiza binti Abd Rashid
Faculty of Applied Sciences, Universiti Teknologi MARA, 40450 Shah Alam, Malaysia

Muhammad Anbariq Abdul Razak
Faculty of Applied Sciences, Universiti Teknologi MARA, 40450 Shah Alam, Malaysia

Nurul Hanani Remeli
Faculty of Electrical Engineering, Universiti Teknologi MARA, 40450 Shah Alam, Malaysia

Budi Aslinie Md Sabri
Centre for Population Oral Health and Clinical Prevention Studies, Faculty of Dentistry, Universiti Teknologi MARA, 40450 Shah Alam, Malaysia

Luqman Saidi
Faculty of Applied Sciences, Universiti Teknologi MARA, 40450 Shah Alam, Malaysia

Sabiha Hanim Saleh
Faculty of Applied Sciences, Universiti Teknologi MARA, 40450 Shah Alam, Malaysia

Ahmad Faiz Mohd Sampian
Faculty of Electrical Engineering, Universiti Teknologi MARA, 40450 Shah Alam, Malaysia

Siti Aisyah binti Shamsuddin
Faculty of Applied Sciences, Universiti Teknologi MARA, 40450 Shah Alam, Malaysia

Mohd Suffian Sulaiman
Faculty of Computer and Mathematical Sciences, Universiti Teknologi MARA, 40450 Shah Alam, Malaysia

Mohd Suhaimi Sulaiman
Faculty of Electrical Engineering, Universiti Teknologi MARA, 40450 Shah Alam, Malaysia

Nordiana Suhada Mohamad Tahiruddin
Faculty of Applied Sciences, Universiti Teknologi MARA, 40450 Shah Alam, Malaysia

Mohd Asmawi Mohd Tayid
Department of Veterinary Services Kuala Langat, Jalan Sungai Buaya, 42600, Sungai Jarum, Kuala Langat, Malaysia

Mazni Mohd Yati
Faculty of Applied Sciences, Universiti Teknologi MARA, 40450 Shah Alam, Malaysia

Farzini Yusha
Faculty of Applied Sciences, Universiti Teknologi MARA, 40450 Shah Alam, Malaysia

Farida Zuraina binti Mohd Yusof
Faculty of Applied Sciences, Universiti Teknologi MARA, 40450 Shah Alam, Malaysia

Habsah binti Zahari
Faculty of Applied Sciences, Universiti Teknologi MARA, 40450 Shah Alam, Malaysia

Muhammad Zaki Zakaria
Faculty of Applied Sciences, Universiti Teknologi MARA, 40450 Shah Alam, Malaysia

Nurul Farhana binti Zulkifli
Faculty of Applied Sciences, Universiti Teknologi MARA, 40450 Shah Alam, Malaysia

Nurulfarhana Zulkifli
Faculty of Applied Sciences, Universiti Teknologi MARA, 40450 Shah Alam, Malaysia

LIST OF ABBREVIATIONS

AChE	acetylcholinesterase
AFR	age at first reproduction
AI	artificial intelligence
ANN	artificial neural network
ANOVA	analysis of variance
ASP	advanced signal processing
BA	betulinic acid
BHA	butylated hydroxyanisole
BLAST	basic local alignment search tool
BSA	bovine serum albumin
CBCT	cone beam computed tomography
CBIR	content-based image retrieval
CDI	1,1-carbonyldiimidazole
CDW	cell dry weight
DFO	1,8-diazafluoren-9-one
DPPH	1,1-diphenyl-2-picrylhydrazyl
EC	egg clutch
EDT	egg development time
E–I	enzyme–inhibitor
EM	effective microorganisms
EMAS	EM activated solution
FEC	fecal egg count
FIS	fuzzy inference system
FL	fuzzy logic
FN	false negative
FP	false positive
FPR	false positive rate
FRIM	Forest Research Institute Malaysia
FTA	fast technology for analysis
FTIR	Fourier transform infrared spectroscopy
GPS	global positioning system
GSR	gunshot residue
Hb	hemoglobin

HCT	hematocrit
HDL	high-density lipoprotein
HF	high frequency
IBS	irritable bowel syndrome
ICS	Image Capturing Studio
IR	image retrieval
IS	impedance spectroscopy
LD	lethal dose
LF	low frequency
LM	Levenberg–Marquardt
MCL	medium chain length
MEOR	microbial-enhanced oil recovery
MIC	minimal inhibitory concentration
MLP	multilayered perceptron
MOF	metal–organic framework
MOHE	Ministry of Higher Education
MRS	de Man, Rogosa, and Sharp medium
MSM	minimal salt medium
NPRF	navigation pattern relevance feedback
NSAID	nonsteroidal anti-inflammatory drugs
OD	optical density
OS	operating system
PCV	packed cell volume
PHA	polyhydroxyalkanoate
PMI	postmortem interval
QBE	query-by-example
QE	query expansion
QPM	query point movement
QR	query reweighting
RBF	radial basis function
RDF	resource description framework
RF	relevance feedback
RFID	radio frequency identification
ROC	receiver operating characteristics
RSM	response surface methodology
SBIR	semantic-based image retrieval
SDK	software development kit
SDS	sodium dodecyl sulfate

SEM	scanning electron microscope
SGF	simulated gastrointestinal fluid
SHF	super high frequency
SIF	simulated intestinal fluid
SINEs	short interspersed elements
SSE	sum square error
TBIR	text-based image retrieval
TG	triglycerides
TN	true negative
TP	true positive
TPR	true positive rate
UHF	ultrahigh frequency
UI	user interface
UiTM	Universiti Teknologi MARA
VIS	visible spectrum

PREFACE

This book focuses on the recent advances and applications in tropical agriculture and bioresources. It outlines some of the very recent advances, basic tools, and the applications of novel approaches to improve agricultural practices and the utilization of bioresources for the enhancement of human life. Highlights of this book are the thorough discussions on various aspects of agricultural modernization through technological advances in information technology, efficient utilization of underexploited natural bioresources, new chemical approaches for the generation of novel biochemicals, and the applications of forensic and genetic approaches for bioresource conservation. The book also walks the readers through the latest advancements in the field of agriculture and applied biology such as the use of radio frequency ID for monitoring agricultural crop and biological applications.

Written by highly experienced and innovative minds in each field of study, this book will help professionals and students understand cutting-edge and futuristic areas of agricultural and biological applications, facilitating further research and application of methods into real-world solutions for some of the most burning questions in sustainable agriculture and natural resource utilization.

INTRODUCTION

This book presents readers with a range of topics that showcase the innovative use of modern technologies and biological principles to improve agriculture practices and enhance the use of underexploited bioresources. It looks at new synthesis approaches for the generation of novel biochemicals. The content presented serves as a source of ideas and inspiration for further innovation of biological applications to enhance human life.

Part I of the book discusses the use of modern information technology in agriculture, such as the use of RFID for livestock monitoring and fuzzy logic to predict fruit ripeness. Part II focuses on biotechnological exploitation of neglected bioresources such as lichens and uncommon bacteria, and also highlights the use of modern biological applications in various forensic and conservation biology. Chapters in Part III provide new insight into innovative approaches for the synthesis of novel biochemicals. Part IV focuses on the synthesis of novel biochemical compounds such as naturally occurring natural product betulinic acid derivatives with amino acids (alanine, phenylalanine, leu-ala) since these compounds have the potential to be anti-cancer, anti-bacteria, and anti-inflammatory.

The synthesis of azidotrimethyl cyclohexene also is discussed because these compounds may serve as the starting material for the synthesis of anti-influenza drugs, e.g, Tamiflu. This part also discusses the synthesis of Schiff base ligands and their physical studies that can be tested on mice. The synthesis of an amino base–metal framework for the application in catalysis is also discussed.

The book covers a wide range of topics, including
- the use of information technology and mathematics in agricultural practices
- utilization of underexploited bioresources
- innovative applications in genetics and forensic biology
- novel approaches for the synthesis of new biochemicals

PART I
Technological Advances for Sustainable Agricultural Practices

CHAPTER 1

A FUZZY APPROACH TO CATEGORIZING RIPENESS OF *CITRUS SUHUENSIS* USING SELECTED OPTICAL INDICES

NOOR AISHAH KHAIRUZZAMAN*, HADZLI HASHIM, NOOR EZAN ABDULLAH, MOHD SUHAIMI SULAIMAN, AZRIE FARIS MOHD AZMI, and AHMAD FAIZ MOHD SAMPIAN

Faculty of Electrical Engineering, Universiti Teknologi MARA, 40450 Shah Alam, Malaysia

Corresponding author. E-mail: noor.aishah88@yahoo.com

CONTENTS

ABSTRACT

This chapter is about applying fuzzy logic to categorize the ripeness of *Citrus suhuensis* using the reflectance measurement. In this study, only selected optical indices were used as reference input to design the fuzzy logic model which are orange, yellow, and green. The wavelength index for yellow is 570 nm, orange at 590 nm, and green at 510 nm. *Citrus suhuiensis* has same color of skin within the growth stage and maturity stage. The reflectance measurement was taken from an external part, the skin of *Citrus suhuiensis*. Each *Citrus suhuiensis* was divided into six portions. The data were obtained by using spectrometer. All the data were analyzed using SPSS software for future analysis. For the statistical analysis, input was chosen for fuzzy system. There were two techniques proposed and experimented in this work termed as direct and differential. Direct technique is a direct reflection measurement of optical index representing orange and yellow. Differential technique takes the reflectance measurement slope of orange and yellow with respect to green indices. At the end, all the 212 samples of *Citrus suhuensis* were successfully tested in this work.

1.1   INTRODUCTION

Limau madu, scientific name *Citrus suhuensis* Hort., belongs to Rutaceae family that has a high potential for development for those people who are interested in upgrading the quality of local fruits [1]. The *Citrus* tree is not too high but shady and its fruits are easy to pluck by hand. Maturity stage of the fruits are between 8 and 10 months [2]. *C. suhuensis* plants do not have a major fruit season, but its production is based on current climatic conditions, terrain, and soil suitability changes [1]. During ripeness or after maturity stage, the size of the fruits is large with normally sweet juicy orange and yellowish content. The contents have less fiber and the fruits are easy to peel [3]. However, externally it is difficult to identify whether the *C. suhuensis* fruit is ripe because the color of the skin does not change a lot which is still green in color. In other words, the color is similar before and after maturity stage. The fruit's content is of whitish dull color, its skin is a little bit dry and rough during unripe stage and later becomes little bit oily and soft when it is ripe. The ability to classify ripeness and the unripeness level of *C. suhuensis* will benefit both farmers and consumers. With this classification, farmers can easily grade the fruits and sell them with a competitive price, optimally in the fruit industry.

There was a lot of research done on studying the ripeness level of local fruits or healthy condition of trees. Most of them used intelligent algorithms such as fuzzy logic (FL) or artificial neural network (ANN) for pattern classifiers. For detection phase, they used image processing for significant features extraction, for example, identification of ripeness level of watermelon [4], papaya [5], tomato [6], and bananas [7]. For the former research, FL was used as the modeling system to process image information that represents ripeness condition for watermelon [8], oil palm fruit [9], and olive [10]. While in the latter research, ANN made use of quantified pixels extracted from image processing technique to indicate the ripeness level [7]. In other work, fuzzy system was also used in classifying rubber tree health condition from white root disease. The system is capable of classifying the tree stage as healthy, medium, or worst [11]. In addition, fuzzy system is useful in grading the apple fruit based on their quality features [12] and in robotic field, fuzzy system is used for controlling the movement of an autonomous mobile robot [13]. Currently, there is no device designed to measure the ripeness and unripeness level of *C. suhuensis*. The standard of procedure still depends on experts to identify the fruit ripeness as well as calculating and recording the fruits' maturity time.

This work has proposed using FL algorithm to classify the ripeness and unripeness of *C. suhuensis*. As far as the detection stage is concerned, optical properties of the fruit skin are measured using spectrometer within the visible spectrum (VIS). Since the fruit contents are orange and yellowish in color during ripeness, the skin's reflectance optical properties in the orange and yellow band spectrum are measured, observed, and analyzed for significant features extraction. These optical data indices are then trained in optimizing fuzzy system. The accuracy is validated at the end stage with a new testing set of measurements representing ripeness and unripeness.

1.2 MATERIALS AND METHODS

Figure 1.1 shows the flowchart for the overall process and steps taken to complete the work. The objective is to determine the ripeness and unripeness of *C. suhuensis* from reflectance measurement and the process is divided into several steps.

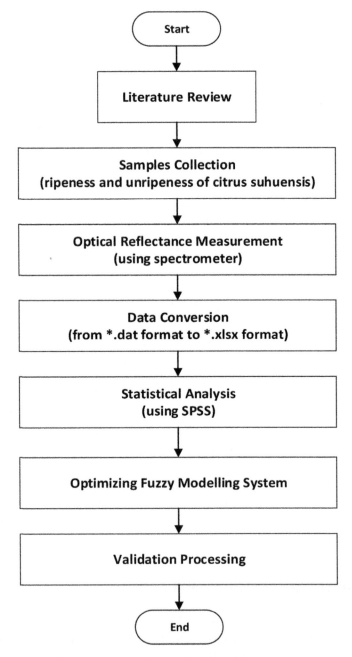

FIGURE 1.1 Flowchart for overall process and steps taken to complete the work.

1.2.1 REFLECTANCE MEASUREMENT

In this work, spectrometer was used to measure the amount of reflectance light and it is operated by transmitting light beam from a source and its intensity is read back by a receiver or detector. Spectrometer used in this work is manufactured by Carl Zeiss with a model number MCS600. The head (OFK30) containing the transmitter and receiver for measuring the spectral range is attached to MCS600 model via an optical cable. This equipment set has the capability to measure percentage of reflectance light from any surface and the spectrum wavelength range covers from 190 to 2200 nm (VIS). Figures 1.2–1.4 show the spectrometer operation. The optical light reflectance measurements can be displayed graphically using Aspect Plus software. Examples of the graphs are shown in Figures 1.5 and 1.6, respectively. This instrument set is located at the Image Capturing Studio (ICS), Advanced Signal Processing (ASP) Lab, Faculty of Electrical Engineering, Universiti Teknologi Mara, Shah Alam.

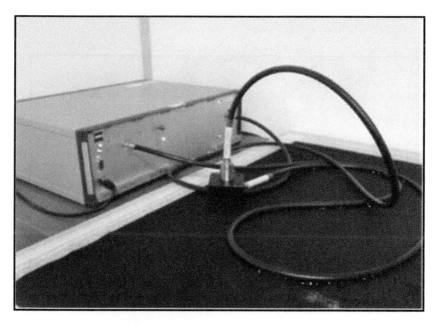

FIGURE 1.2   Spectrometer.

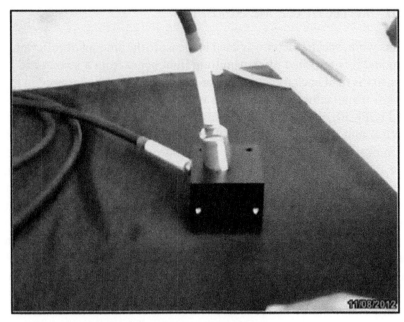

FIGURE 1.3   The OFK30 type attached with spectrometer.

FIGURE 1.4   The OFK30 type attached to the sample.

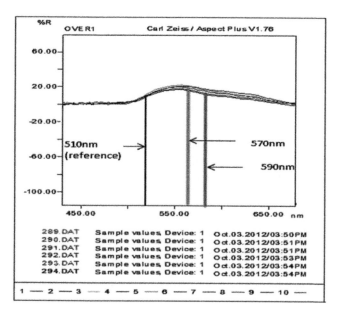

FIGURE 1.5   Direct technique: only yellow (570 nm) and orange (590 nm) wavelength measurements are considered.

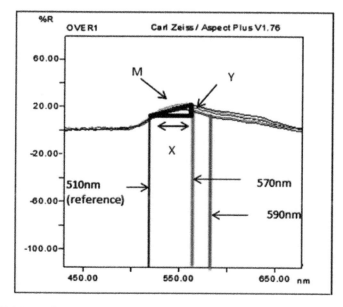

FIGURE 1.6   Differential technique: yellow (570 nm) and orange (590 nm) wavelength measurements are referred to green (510 nm) by finding their slope indices.

There were two techniques proposed and experimented in this work. They are termed as direct and differential technique where the former means any reflection measurements from the spectrometer are taken directly from the sample with respect to orange (590 nm) and yellow (570 nm) color wavelength, respectively, as shown in Figure 1.5. The latter differential technique is where gradient slope indices are calculated by taking reflection measurements of orange and yellow color wavelength with respect to the green wavelength (510 nm) as illustrated in Figure 1.6. The rationale for taking green wavelength as reference is that it represents the fruit's skin color during ripeness or unripeness stage. At the classification stage, measurements from both techniques were used to design an FL model and the optimized system was validated with testing set of data. The outcome would provide the best model for classifying ripeness or unripeness of *C. suhuensis*.

With respect to Figure 1.6, here is an example to find a slope index

$$M = \frac{Y}{X} \, , \tag{1.1}$$

where
M = slope for ripe level,
Y = reflectance measurement of yellow/orange (%) minus reflectance measurement of green (%),
X = wavelength of yellow/orange (nm) minus wavelength of green (nm).

1.2.2 DETECTION STAGE

Reflectance measurements from Aspect Plus software is in *.DAT format. Since measurements need to be analyzed statistically using SPSS analysis, all data saved in Aspect Plus were converted into *.xlsx format and stored in Microsoft Excel. The data were later transferred into SPSS for descriptive analysis. This software was used to compare the reflectance measurements for direct and differential technique in terms of boxplot.

1.2.3 CLASSIFICATION STAGE

The analyzed output from SPSS would indicate features that could be used as input in training and optimizing an FL model for classifying ripeness or unripeness of *C. suhuensis*. For developing the fuzzy system, five elements needed to be developed in the fuzzy inference system (FIS). They are the FIS editor, membership function editor, rule editor, rule viewer, and surface viewer [15]. Explanations of the process are described in the next chapter.

1.3 RESULTS AND DISCUSSION

In this work, SPSS provides boxplots information about the minimum, mean, and maximum value for each data sample of reflectance representing ripeness and unripeness cluster group for both techniques. Figures 1.7 and 1.8 illustrate the location of the measured data. The middle quantile is the location of the box in the plots while the straight lines are for the upper and lower quantiles. From statistical knowledge, these boxes demonstrate the location of 50% of the measured data. The upper and lower lines represent another 25% each, respectively. In addition, location of outliers is also shown which are normally scattered beyond the boxplot.

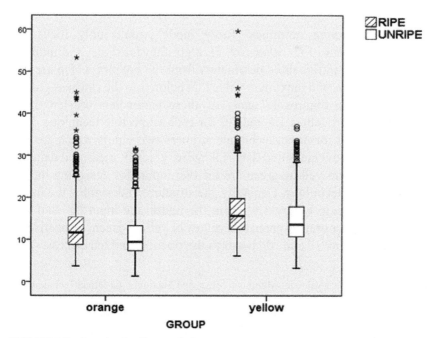

FIGURE 1.7   Boxplot for direct technique.

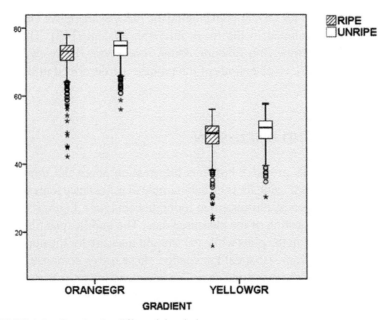

FIGURE 1.8   Boxplot for differential technique.

Since designing optimized fuzzy model would only focus on detecting ripeness of *C. suhuensis*, FL algorithm for direct and indirect technique shall utilize three parameters from the boxplot which are the minimum, mean, and maximum value that belongs to the ripeness group. With respect to Figures 1.7 and 1.8, these parameters are identified and presented in Tables 1.1 and 1.2 for each respective technique. The designed model for direct technique requires two inputs which are the orange and yellow measured data. Likewise, gradient measured data for orange-green and yellow-green are the two inputs for designing model for differential technique. Generally, the structure in designing the fuzzy model is shown in Figure 1.9. From the figure, the input "*o*" and "*y*" depicts orange or orange-green and yellow or yellow-green, respectively. While the output "1" and "0" portrays the ripeness and unripeness.

TABLE 1.1   Selected Value for Minimum, Mean, and Maximum for Direct Technique.

Group	Orange			Yellow		
	Min	**Mean**	**Max**	**Min**	**Mean**	**Max**
Ripe	3.71	13.00	53.26	6.16	16.94	59.52

TABLE 1.2   Selected Value for Minimum, Mean, and Maximum for Differential Technique.

Group	Orange-green			Yellow-green		
	Min	**Mean**	**Max**	**Min**	**Mean**	**Max**
Ripe	42.18	71.95	78.09	15.92	47.97	56.04

The FL toolbox includes an FIS that involves all elements that were described in the section above [11]. The process of FIS can be followed using the FIS editor displayed from MATLAB software. Figures 1.10 and 1.11 show the FIS editor for direct and differential technique, respectively.

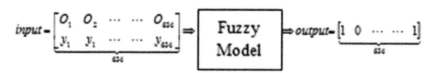

$$input = \begin{bmatrix} O_1 & O_2 & \cdots & \cdots & O_{634} \\ y_1 & y_1 & \cdots & \cdots & y_{634} \end{bmatrix}_{634} \Rightarrow \boxed{\begin{array}{c} \textbf{Fuzzy} \\ \textbf{Model} \end{array}} \Rightarrow output = \underbrace{\begin{bmatrix} 1 & 0 & \cdots & \cdots & 1 \end{bmatrix}}_{634}$$

FIGURE 1.9   Structure for designing the fuzzy model.

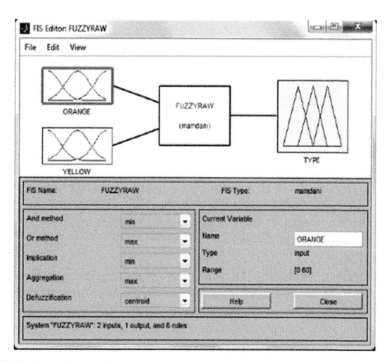

FIGURE 1.10   Fuzzy inference system (FIS) editor for direct technique.

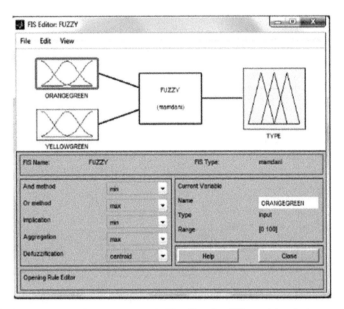

FIGURE 1.11   Fuzzy inference system (FIS) editor for differential technique.

The values of minimum, mean, and maximum were inserted at the input in the membership function editor and assigned as parameters, while the output variable at the fuzzy toolbox was set as "RIPE" and "UNRIPE." The examples of membership function editor for orange and orange-green parameter input are shown in Figures 1.12 and 1.13, respectively.

Rule viewer shows the result of the FIS at the end of the process which used to be for verification of the whole work. As an example and with respect to Figure 1.14, the process started when the orange was at 4.25 and yellow at 7.28. Then, the result shows that it is categorized as unripe with 0.394.

Figure 1.15 is another example as in the case of differential technique using the rule viewer. It shows that when the gradient orange-green is at 49.1 and gradient yellow-green is at 28.4, the result indicates that it is unripe with 0.3.

Altogether, a total of 1268 data points comprising orange and yellow data sets were verified in the fuzzy system in the MATLAB software. In this work, sensitivity is used to measure the percentage of true positive where samples are correctly identified as ripe, while specificity is used to measure the percentage of true negative that correctly identified as unripe. Table 1.3 shows the result of sensitivity and specificity for direct technique.

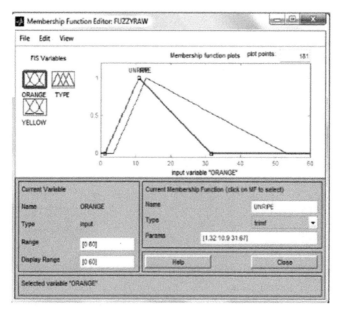

FIGURE 1.12   Membership function editor for orange parameter input.

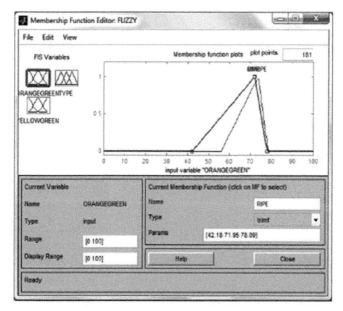

FIGURE 1.13   Membership function editor for gradient orange-green parameter input.

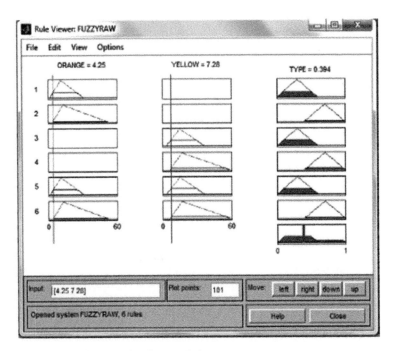

FIGURE 1.14   Rule viewer for direct technique.

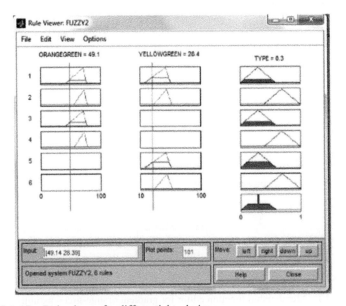

FIGURE 1.15   Rule viewer for differential technique.

TABLE 1.3 Result for Sensitivity and Specificity for Direct Technique.

		Actual	
		Ripe	Unripe
Predicted	Ripe	330	230
	Unripe	304	404
		Sensitivity	Specificity
		$= \dfrac{330}{634} \times 100 = 52.05\%$	$= \dfrac{404}{634} \times 100 = 63.72\%$

From the result of sensitivity and specificity, it is shown that the reflectance of orange and yellow can categorize the ripe (the sensitivity) and unripe (the specificity) with 52.05% and 63.72%, respectively. The outcome implies that the developed fuzzy model can recognize 330 of ripe and 404 of unripe using the spectrum color of orange and yellow. Therefore, the overall accuracy is calculated as shown below.

$$\text{Accuracy} = \frac{330 + 404}{634 + 634} \times 100 = 57.9\% . \qquad (1.2)$$

Table 1.4 describes the result for sensitivity and specificity of ripe and unripe in differential technique.

TABLE 1.4 Result for Sensitivity and Specificity for Differential Technique.

		Actual	
		Ripe	Unripe
Predicted	Ripe	359	449
	Unripe	275	185
		Sensitivity	Specificity
		$= \dfrac{359}{634} \times 100 = 56.62\%$	$= \dfrac{185}{634} \times 100 = 29.18\%$

With respect to the table, the results of using reflectance measurement with differential technique could categorize ripeness (the sensitivity) of 56.62% and unripeness (the specificity) of 29.18%. Using this technique, the fuzzy model does not have the capability to recognize unripeness of fruit and thus, reducing its overall accuracy in the detection as shown below.

$$\text{Accuracy} = \frac{359+185}{634+634} \times 100 = 42.9\% .$$ (1.3)

1.4 CONCLUSION

The maturity stage of *C. suhuensis* can be divided into ripeness and unripeness. In this research, the optical reflectance properties in the VIS of 212 *C. suhuensis* fruits were being measured using spectrometer. From the measured reflectance, 634 samples were recorded, each set representing orange and yellow wavelength. These colors were chosen because it emulates ripeness of the fruit. In addition, 634 samples of green color wavelength were also recorded representing color of the fruit skin. Two sets of techniques were proposed; the first is known as direct technique where measurements at orange and yellow wavelength were used directly in the process. Meanwhile, the second technique is called differential technique where gradient or slope was calculated for each measurement at orange and yellow wavelength with respect to the green wavelength. These data sets were later analyzed using SPSS software. The generated boxplots from the software have given measurement readings representing the maximum, mean, and minimum for each cluster group.

Only the measurements representing ripeness were applied as input in designing a fuzzy model for detecting ripeness. They were inserted in the membership function editor and assigned as parameters, while the output variable was set as "RIPE" and "UNRIPE". After validating with 634 samples for each ripeness class, the fuzzy model for direct technique produced an overall accuracy of 57.9%, with a sensitivity of 52.05% and specificity of 63.72%. As for differential technique, the overall accuracy is calculated as 42.90%. This model's sensitivity and specificity is 56.62% and 29.18%, respectively. It can be concluded that both fuzzy models produced poor performance in detection with low overall accuracy. However, direct technique model has advantage over the differential method as its specificity is much higher. Indirectly, it can detect proportionally if given a set of ripe and unripe *C. suhuensis*.

ACKNOWLEDGMENT

The authors would like to thank Mr. Hussein Bin Abdullah, an expert from the Limau Madu farm at Jeli, Kelantan, from where the *Citrus suhuensis* fruits were collected. Our greatest gratitude also goes to UiTM Research Management Institute (RMI) and Faculty of Electrical Engineering, Universiti Teknologi MARA for supporting this work under Excellent Fund Grant (Research Intensive Faculty) under the code (81/2012).

KEYWORDS

- *Citrus suhuensis*
- ripeness
- color
- fuzzy approach
- optical indices

REFERENCES

1. Haiman. Projek Pertanian. http://tipsprojektani.blogspot.com/2011/02/penjagaan-limau-madu.html (accessed Feb 24, 2011).
2. Manjung, P. P. D. In Panduan Menanam Limau (Tanaman Jangka Panjang), ed: Disember 2007–2010. http://pertanianmjg.perak.gov.my/bahasa/panduan_tanamlimau.htm.
3. Yusof, Y. W. M.; Hashim, H.; Noh, N. B. M.; Abdullah, N. E.; Osman, F. N. A Preliminary Study of Citrus Suhuensis Flavor Based on RGB Color Information. *2nd International Conf on Intell. Syst. Model. Simul.* Phnomh Penh, 2011, pp 100–105.
4. Abdullah, M. M.; Abdullah, N. E.; Hashim, H.; Rahim, A. A. A.; George, C.; Igol, F. A. In IEEE Student Conference on Research and Development (SCOReD), Various Grades of Red Flesh Watermelon Ripeness Based on NIR and VIS Reflectance Measurement, Penang, 2012, pp. 250–255.
5. Saad, H.; Hussain, A. In 4th Student Conference on Research and Development (SCOReD), Classification for the Ripeness of Papayas Using Artificial Neural Network (ANN) and Threshold Rule, Shah Alam, 2006, pp. 132–136.
6. Shibata, T.; Iwao, K.; Takano, T. Evaluating Tomato Ripeness Using a Neural Network. *J. Soc. High Technol. Agric.* 1996, *8*, 60.

7. Saad, H.; Ismail, A. P.; Othman, N.; Jusoh, M. H.; Naim, N. F.; Ahmad, N. A. In IEEE International Conference on Signal and Image Processing Applications (ICSIPA), Recognizing the Ripeness of Bananas Using Artificial Neural Network Based on Histogram Approach, Kuala Lumpur, 2009, pp. 536–541.

8. Rahman, F. Y. A.; Baki, S. R. M. S.; Yassin, A. I. M.; Tahir, N. M.; Ishak, W. I. W. In WRI World Congress on Computer Science and Information Engineering, Monitoring of Watermelon Ripeness Based on Fuzzy Logic, Los Angeles, 2009, pp. 67–70.

9. May, Z.; Amaran, M. In Proceedings of the 13th WSEAS International Conference on Mathematical and Computational Methods in Science and Engineering (MACMESE 2011), Automated Ripeness Assessment of Oil Palm Fruit Using RGB and Fuzzy Logic Technique, Wisconsin, USA, 2011, pp. 52–59.

10. Aparicio, R.; Morales, M. T. Characterization of Olive Ripeness by Green Aroma Compounds of Virgin Olive Oil. *J. Agric. Food Chem.* **1998**, *46*, 1116–1122.

11. Latib, S. B. In International Conference On Electrical, Electronics and System Engineering (ICEESE2013), Fuzzification of Rubber Tree White Root Disease Based on Leaf's Discolouration, Kuala Lumpur, 2013.

12. Kavdir, I.; Guyer, D. E. Apple Grading Using Fuzzy Logic. *Turk. J. Agric. For.* **2004**, *27*, 375–382.

13. Surmann, H.; Huser, J.; Peters, L. In *Fuzzy Systems*, Proceedings of International Joint Conference of the Fourth IEEE International Conference on Fuzzy Systems and The Second International Fuzzy Engineering Symposium. A Fuzzy System for Indoor Mobile Robot Navigation, Yokohama, 1995, vol. 1, pp. 83–88.

14. D. U. Simon, W. Nörthemann, D. B. Ohnesorge, and D. F. Stietz, "spectrometer systems" in *Microscopy–Optical Sensor Systems*, ed. Germany: Carl Zeiss Microscopy GmbH.

15. Adel Abdennour, E. E. D.; K. S. University. Tutorial on Fuzzy Logic Using MATLAB. p. 18, 16 November 2012.

CHAPTER 2

AUTOMATED VISIBLE COLOR SPECTRUM MODEL FOR RECOGNIZING RIPENESS OF *CITRUS SUHUIENSIS*

FARIDATUL AIMA ISMAIL*, HADZLI HASHIM,
NINA KORLINA MADZHI, NOOR EZAN ABDULLAH, and
NURUL HANANI REMELI

Faculty of Electrical Engineering, Universiti Teknologi MARA (UiTM) Selangor, Malaysia

Corresponding author. E-mail: aima_ismail@yahoo.com

CONTENTS

ABSTRACT

The aim of this study is to implement an artificial neural network (ANN) technique in order to differentiate between ripeness and unripeness stages of *Citrus suhuiensis*. Initially, these stages will be measured optically using nondestructive method via spectrometer of MSC600 Carl Zeiss. The spectrometer will transmit VIS (visible spectrum) photonic light radiation to the surface (skin) of the sample. The reflected light from the sample's surface will be received and measured by the same spectrometer in terms of percentage that will form a line pattern representing each wavelength component in the VIS range. These patterns of wavelength components will be used as input parameters for designing two optimal ANN classifiers, where each one of them will be supervised with Levenberg–Marquardt (LM) and radius basis function (RBF) algorithm, respectively. The performance of the optimized model is decided after observing the receiver operating curve (ROC) plot. The result outcomes have shown the optimized LM trained model has better performance in terms of sensitivity, specificity, and accuracy, and outclassed previous intelligent identification models when validated at a threshold of ± 0.4. The overall accuracy for LM algorithm is 71.5% whilst the true positive rate (TPR) of each case is 54% for ripe and 89% for unripe.

2.1 INTRODUCTION

Limau madu or *Citrus suhuiensis* Hort. ex.Tanaka [1] is a nonseasonal fruit that can be found mostly in East Malaysia area, especially in Kelantan. The stage of fruit maturity is an important aspect because it can determine the right time for harvesting the fruit [2]. Usual practice of maturity or ripeness determination of fruits is by monitoring several factors either using destructive or nondestructive method based on its skin color, fruit size, flesh color, number of days after planting, and other relevant criteria [3]. *C. suhuiensis* has a sweet and sour taste, contains high juice percentage, various size conditions, and has less fiber and is favorable. However, this fruit has a unique feature which is the skin color is green from early growth until ripening stage. Due to this special feature, the cultivators may face problems in differentiating between under ripe and ripe [4] based on its skin color unless they chose to peel of the skin and taste it. With regard to the issues, hence, the purpose solution is to study the nondestructive technique on the maturity stage identification of the *C. suhuiensis*. It is important to know the stage of maturity for determining when to harvest fruit since fruit harvested at an immature stage will not be able to achieve a level of quality acceptable to consumers. It also has a different color and taste. Most of the people will prefer sweet flavor more than other flavor [4].

The main objective of this study is to categorize ripe and unripe of *C. suhuiensis* using visible spectrum optical reflectance measurement. While another objective is to design an artificial intelligence classification model by utilizing measured reflectance's pattern as the input features. There are several studies which were carried out on determination of fruit ripeness and color feature determination using optical properties via spectrometer, such as on watermelon ripeness [5], rubber tree leaf diseases [6], white root diseases [7], and rubber seeds [8]. From the findings, it shows that the spectrometer is able to extract useful information through color spectrum and suitable for determination of ripening stage of citrus via nondestructive technique. While the well-known method of classification system utilizing the artificial neural network (ANN) as the classifier system has been employed in agricultural studies such as in classification of rubber tree leaf disease [9], rubber tree clones via seeds [8], watermelon ripeness [10], and white root diseases [7].

Human eyes are sensitive to light which lies in a very small region of the electromagnetic spectrum labeled "visible light." The wavelength range is around 400–700 nm. The human eye is not capable of "seeing" radiation with wavelengths outside the visible spectrum. This visual spectrum was used in order to see through the sample experiments and it can be related to the electrical instrument that is to be used. This wavelength will be used by spectrometer in order to collect data in term of percentage when reflected light is received back to the spectrometer [11].

ANN was selected as the classifier in developing the intelligent system model for this work. An ANN is configured for a specific application, such as pattern recognition or data classification, through a learning process. Learning in biological systems involves adjustments to the synaptic connections that exist between the neurons [12].

2.2   MATERIALS AND METHODS

The flowchart in Figure 2.1 below shows general phases in determining the ripeness level of *C. suhuiensis*. The first phase is to gather relevant knowledge via literature review. The second phase is process of sample collection of the citrus from the farm in Kelantan. The next phase is developing process of ANN model using Levenberg–Marquardt (LM) and radial basis function (RBF). The final phase is the comparison analysis from both models.

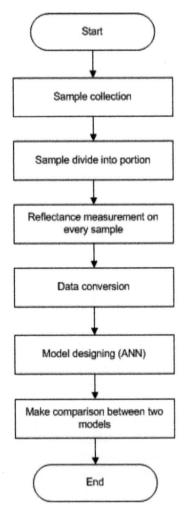

FIGURE 2.1   General phases in classification of *C. suhuiensis*.

2.2.1 CITRUS SUHUIENSIS HORT. EX.TAN SAMPLE

The sample of *C. suhuiensis* Hort. ex.tanaka (limau madu) for this project was bought from farm in Kelantan, which consisted of both ripe and unripe. The selection was done by an expert worker from the farm. A total of 200 *C. suhuiensis* samples have been taken for this project, whereas 100 for ripe samples and another 100 for unripe samples. The harvesting period for unripe sample was done in the middle of October where at this time the flesh color of *C. suhuiensis* is slightly yellowish and the flavor is sour. Meanwhile, the ripe *C. suhuiensis* was taken in the middle of November. The flesh color during ripe stage is orange and the flavor is sweet as shown in Figure 2.2 (A). Each one of the sample is divided in six portions and the reflectance data have been taken in each portions. In addition, each of the samples will be divided into portions so that each characteristic of the different portions can be determined. From these samples, a database comprising 30 inputs has been built representing green, yellow, and orange color spectrum.

(a)   (b)

(c)

FIGURE 2.2 The difference between ripe (A) and unripe (B) flesh color (C) outside appearance of citrus.

2.2.2 REFLECTANCE MEASUREMENT

Zeiss spectrometer is used to measure the reflectance percentage of green, yellow, and orange color from *C. suhuiensis* samples. Visible spectrum (VIS) has been chosen as the light reflectance, where it consists of a spectrum of wavelengths which range from ~700 nm to 400 nm [11] as shown in Figure 2.3. A total of 200 samples of citrus have been taken which are 100 samples for ripe and 100 for unripe. Each citrus are divided into six portions; thus the total experiment data that have been collected are 1200. The reflectance process was done at the Advanced Signal Processing (ASP) Lab, Faculty of Electrical Engineering, UiTM, Shah Alam. The scanned light reflectance information was directly extracted using Aspect Plus software. All these data will be transferred to Microsoft Office Excel later for ease of analysis. The samples are measured in uncut conditions as shown in Figure 2.4. The results obtained will be compared.

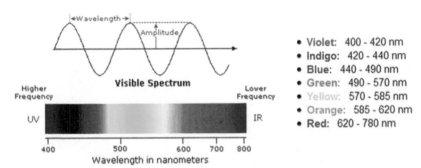

- Violet: 400 - 420 nm
- Indigo: 420 - 440 nm
- Blue: 440 - 490 nm
- Green: 490 - 570 nm
- Yellow: 570 - 585 nm
- Orange: 585 - 620 nm
- Red: 620 - 780 nm

FIGURE 2.3   Optical properties of VIS.

FIGURE 2.4   Uncut condition.

2.2.3 MODEL DESIGNING

The ANN model is developed to differentiate between ripe and unripe *C. suhuiensis*. For the ANN model, multilayered perceptron (MLP) network with one hidden layer was recommended in diagnosing these cases based on the fact that it has been widely applied by many researchers [13]. In addition, there are two network that suits with this diagnosing data, first is LM algorithm with one hidden layer and RBF.

Optimization of the trained models was decided using a confusion matrix [14]. Confusion matrix is a matrix for a two-class classifier, which contains information about actual and predicted classifications done by a classification system [15]. Table 2.1 describes the confusion matrix for a two cases classifier.

TABLE 2.1 Confusion Table.

		Actual Class	
		A+	A−
Predicted class	P+	TP	FP
	P−	FN	TN

The entries in the confusion matrix have the following meaning defined as:

P+ (predicted positive)—the correct predictions
P− (predicted negative)—the incorrect predictions
A+ (actual positive)—the actual correct number
A− (actual negative)—the actual incorrect number
TN (true negative)—the number of correct predictions that an instance is negative
FP (false positive)—the number of incorrect predictions that an instance is positive
FN (false negative)—the number of incorrect predictions that an instance is negative

Every notation in the confusion table is defined as follows:

P+ (**predicted positive**)—the correct predictions
P− (**predicted negative**)—the incorrect predictions
A+ (**actual positive**)—the actual correct number
A− (**actual negative**)—the actual incorrect number
TN (**true negative**)—the number of **correct** predictions that an instance is **negative**
FP (**false positive**)—the number of **incorrect** predictions that an instance is **positive**
FN (**false negative**)—the number of **incorrect** of predictions that an instance **negative**

Table 2.2 tabulates the set of training and testing data that were used in this project. The outputs of the model are the classification of ripe and unripe case and were defined as 10 and 01, respectively.

TABLE 2.2 Proposed Set of Training and Testing Data for Each Classes and Ann Output.

Classes	Training	Testing	Valid	Output
Ripe	400	100	100	10
Unripe	400	100	100	01

2.3 RESULTS AND DISCUSSION

2.3.1 REFLECTANCE MEASUREMENT

Aspect Plus software is then used as a medium to collect data from the light reflectance process of spectrometer. The data are showed as wavelength versus reflectance percentage, where wavelength is represented at x-axis while reflectance percentage is represented at y-axis. Figure 2.5 shows an example when comparing reflectance of spectrum for both stages, ripe and unripe. In this case, only 509–606 nm wavelength ranges have been chosen

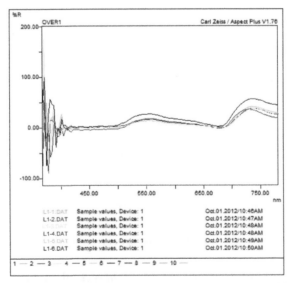

FIGURE 2.5 Comparison of spectrum reflectance of *Citrus suhuiensis* discoloration.

throughout the analysis since the range represents the color spectrum for green, yellow, and orange. Figure 2.5 depicts the overlay of reflectance plots for citrus samples. From this plot, it can be seen that each sample may have slightly different form but they produce the same pattern. The red, blue, violet, and indigo color of wavelength will be eliminated during the analysis because these colors are not related to the prime color of the sample which is orange, yellow, and green.

2.3.2 ANN MODEL DESIGNING

The focus of training model is to differentiate ripe and unripe case from the given training set. The trained model would then be validated using the testing set for performance analysis as being described in Table 2.2. There were only 10 models designed with different number of neurons in each hidden layer which are 3, 7, 15, 18, 21, 25, 32, 35, 44, and 50. The input has 30 nodes representing the range of green, yellow, and orange wavelength spectrum.

Figure 2.6 describes the sum square error (SSE) performance for each designed model versus 100 epochs. All models started to produce converging error during training phase beginning at one epoch cycles. Figure 2.7 illustrates the RBF results at 21 epochs.

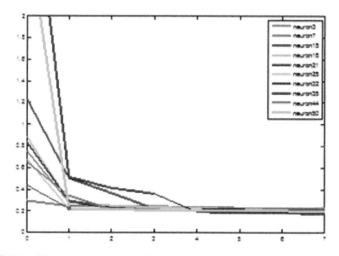

FIGURE 2.6   The sum square error (SSE) versus epoch.

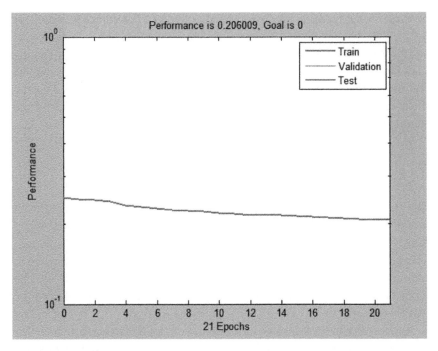

FIGURE 2.7   The performance result for RBF when 21 epochs are applied.

Figure 2.8 shows the percentage accuracy performance calculated from the available confusion matrix for each model. The best hidden layer size would have the highest accuracy and for this LM optimized model, the number of neurons is depicted at 18 (71.5%), while for RBF the highest accuracy is 69% at 21 number of neurons. This result implies that the model is able to recognize evenly for both ripe and unripe.

Equation (1) shows the total number of connections that have been used when designing the ANN model. Thus, the above LM models (18 neurons in the hidden layer) would then produce 606 numbers of connections and for RBF models (21 neurons in hidden layer) produce 702 connections as shown below.

No. of connection = input + (input + hidden size)   (1)
+ (hidden size + output)

No. of connection $_{18}$ (LM) = 30 + (30×18) + (18 +18)
$$= 606$$

No. of connection $_{21}$ (RBF) = 30 + (30n21) + (21+ 21)

$$= 702$$

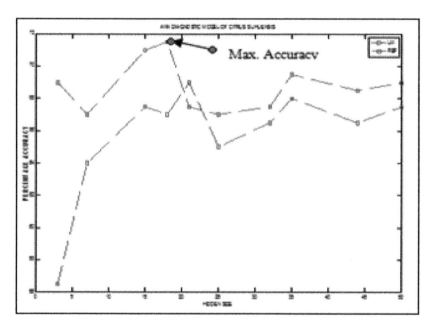

FIGURE 2.8 ANN model showing the 18 neurons are selected because it has the highest accuracy (71.5%).

From the results, the selection of the optimum performance were utilizing the optimized model which is LM model with hidden layer 18 neurons due to its highest accuracy compared to the RBF model. Table 2.3 shows the confusion table of LM model representing the performance of true positive rates (TPRs) for ripe and unripe condition (54% and 89%, respectively). Meanwhile, Table 2.5 shows confusion matrix for RBF model where performance for TPR is 52% for ripe and 86% for unripe.

$$\text{TPR}_{\text{ripe}} \text{LM} = \frac{54}{54 + 46} \times 100\% = 54\%$$

$$\text{TPR}_{\text{unripe}} \text{LM} = \frac{89}{89 + 11} \times 100\% = 89\%$$

$$\text{TPR}_{\text{ripe}} \text{ RBF} = \frac{52}{52 + 48} \times 100\% = 52\%$$

$$\text{TPR}_{\text{unripe}} \text{ RBF} = \frac{86}{86 + 14} \times 100\% = 86\%.$$

The cell (1, 1) and cell (2, 2) of the table show the result of true positive (TP) test applied for each case, 54 representing ripe and 89 unripe. Meanwhile, the cell (3, 3) represents the overall percentage accuracy of 71.5% (Table 2.3) and summarized in Table 2.4.

TABLE 2.3   Confusion Matrix for Optimized LM Algorithm Model with Hidden Layer of 18 Neurons.

		Testing Confusion Matrix		
	0	89	46	65.9%
		44.5%	23.0%	34.1%
Output class	1	11	54	83.1%
		5.5%	27.0%	16.9%
		89.0	54.0%	71.5%
		11.0%	46.0%	28.5%
		0	1	
		Target class		

TABLE 2.4   Confusion Matrix for Ripe for LM Model.

		Actual Case	
		A+	A−
Predicted case	P+	54	11
		(TP)	(FP)
	P−	46	89
		(FN)	(TN)

From the table, the TP, FP, FN, and TN rates are calculated as:

$$\text{TP}_{\text{ripe}} = \frac{54}{54 + 46} (100) = 54\%$$

$$FP_{ripe} = \frac{11}{11+89}(100) = 11\%$$

$$FN_{ripe} = \frac{46}{46+54}(100) = 46\%$$

$$TN_{ripe} = \frac{89}{89+11}(100) = 89\%.$$

TABLE 2.5 Confusion Matrix for RBF Optimized Model with Hidden Layer of 21 Neurons.

		Testing Confusion Matrix		
	0	86	48	64.2%
		43.0%	24.0%	35.8%
Output class	1	14	52	78.8%
		7.0%	26.0%	21.2%
		86.0	52.0%	69.0%
		14.0%	48.0%	31.0%
		0	1	
		Target class		

Receiver operating characteristics (ROC) usually is used to show the TPR versus the false positive rate (FPR) across multiple thresholds. It is used to verify the optimal threshold and to maximize classification accuracy and minimize classification errors [16]. Figures 2.9 and 2.10 show the ROC graph for LM and RBF model. From the graph, the best threshold is 0.4 for both models, whereby at that point the value for FPR is low and the value for TPR is high.

Table 2.6 showed the comparison between all the case optimized models measured at a threshold of +0.4. From the table, the RBF model has the highest number of connection (702) due to greater hidden layer size. However, it has lower balanced percentage performance in terms of sensitivity, specificity, and accuracy. From the table, it is shown that both models, that is, LM model and RBF model are able to recognize between ripe and unripe condition with percentage of total accuracy at 71.5% and 69%, respectively. The specificity for both models is also high with more than 85%. It implies that the model is easy to recognize unripe more than

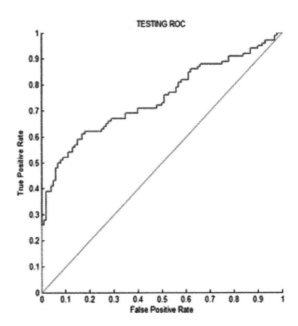

FIGURE 2.9   ROC graph of neurons 18 for LM model.

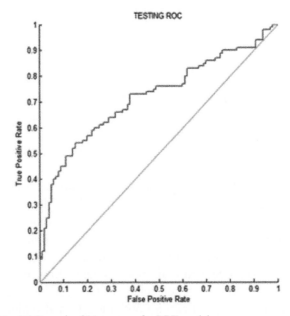

FIGURE 2.10   ROC graph of 21 neurons for RBF model.

ripe condition. In RBF case, even the accuracy is almost similar with LM model but it has disadvantage where the model needs 702 number of connections, which indicates higher cost production if goes for development. If the accuracy and the cost are being considered as the main factors for choosing the model, then LM model is chosen as the best because of low cost (due to low number of connections) and high accuracy compared with RBF model.

TABLE 2.6 Comparison Between Ann Models for this Case.

Model Performance Indicators at Threshold ± 0.4			
Model	LM		RBF
ANN (Input: Hidden: Output)	Ripe	Unripe	Ripe	Unripe
	30:18:01		30:21:01	
ANN (no of connection)	606		702	
Total accuracy (%)	71.5%		69%	
Sensitivity (%)	54%		52%	
Specificity (%)	89%		86%	

2.4 CONCLUSION

In this work, an experiment was conducted to investigate two types of maturity stage of *C. suhuiensis* which are ripe and unripe with a classifier system using two ANN models which are LM model and RBF model. The sensitivity result for both models is around 52–54%, whereas the specificity is at 86–89%. These results imply that they are both strong in classifying unripe stage (higher in specificity) of *C. suhuiensis* compared to ripe stage. As an overall, both ANN model produce overall accuracy more than 65% with LM model having the highest accuracy compared to RBF model which indicate that the classification system is success.

2.5 ACKNOWLEDGMENT

I would like to express my thanks to Faculty of Electrical Engineering UiTM, Research Management Institute (RMI) UiTM Shah Alam and Excellent Fund Grant (Research Intensitive Faculty) code 44/2012 for funding this project.

KEYWORDS

- *Citrus suhuiensis*
- ANN
- RBF model
- LM model
- unripe

REFERENCES

1. Sule, H., et al. Dipersion Pattern and Sampling of Diaphorina citri Kuwayama (Hemiptera:Psylidae) Populations on *C. suhuiensis* Hort. ex Tanaka in Padang Ipoh Terengganu, Malaysia. *Pertanika J. Trop. Agric. Sci.* **2012,** *35*(S), 25–36.
2. Panduan menanam limau. Available from: http://www.pertanianperak.gov.my (accessed April 2013).
3. Slaughter, D.C. *Nondestructive Maturity Assessment Methods for Mango: A Review of Literature and Identification of Future Research Needs;* National Mango Board: Orlando, FL, USA, 2009; pp 1–18.
4. Yuslinda Wati Mohamad Yusof, H. H.; Norhafizah bte Mohd Noh; Noor Ezan Abdullah and a. F. N. Osman. *A Preliminary Study of Citrus Suhuensis Flavor Based on RGB Color Information;* Faculty of Electical Engineering UiTM: Shah Alam, 2012.
5. Abdullah, M.M., et al. *Various Grades of Red Flesh Watermelon Ripeness based on NIR and VIS Reflectance Measurement,* in *Student Conference and Research Development (SCOReD 2012)*December 2012: Batu Feringhi Penang Malaysia.
6. Hashim, H., et al. In *Classification of Rubber Tree Leaf Disease Using Spectrometer,* 2010 Fourth Asia International Conference on Mathemathical/Analytical Modelling and Computer Simulation, 2010.
7. Rahmat, S. S. B. *Early Detection of White-root Disease for Rubber Tree Based on Leaf Discoloration with Neural Network Technique;* Bachelor of Electrical Engineering, FKE, UiTM: Shah Alam, 2012.
8. Hashim, H., et al. In *An Intelligent Classification Model for Rubber Seed Clones Based on Shape Features Through Imaging Techniques,* International Conference on Intelligent Systems, Modelling and Simulation, Liverpool, United Kingdom, 2010, 25–31.
9. Abdullah, N. E., et al. In *Classification of Rubber Tree Leaf Diseases using Multilayer Perceptron Neural Network,* 5th Student Conference and Research Development (SCOReD 2007), Malaysia, Dec 11–12, 2007.
10. Kutty, S. B., et al. In *Classification of Watermelon Leaf Diseases Using Neural Network Analysis,* 2013 IEEE Business Engineering and Industrial Applications Colloquium (BEIAC 2013) Bayview Hotel Langkawi, Malaysia, 459–464.

11. Bruno, T. J.; Svoronos, P. D. N. *CRC Handbook of Fundamental Spectroscopic Correlation Charts;* CRC Press and Taylor & Francis Group: Boca Raton, FL, 2005; pp 1–16.
12. Maria-Sanchez, P. *Neuronal Risk Assessment System for Construction Projects*; Expert Verlag, 2005; 1–2.
13. Jain, A. K.; Mohiuddin, K. M. Artificial Neural Network: A Tutorial, Michigan State University, IBM Almedan project Center, March 1996.
14. Visa, S., et al., *Confusion Matrix-based Feature Selection.*
15. Krogh, A. What Are Artificial Neural Networks? *Nat. Biotechnol.* **2008,** *26*(2), 195–197.
16. T. F. (tom.fawcett@hp.com) and M. HP Laboratories, 1501 Page Mill Road, Palo Alto, CA 94304, ROC Graphs: Notes and Practical Considerations for Researchers, March 16, 2004.

CHAPTER 3

EFFECT OF TEMPERATURE ON THE LIFE PERFORMANCE OF TROPICAL CLADOCERAN, *CERIODAPHNIA CORNUTA* FROM TASIK ELHAM, PERLIS

ROSLINDA BINTI ISMAIL[1], HASNUN NITA BINTI ISMAIL[1,2,*], and TAY CHIA CHAY[1,3]

[1]*Faculty of Applied Sciences, Universiti Teknologi MARA, 40450 Shah Alam, Malaysia*

[2]*Faculty of Applied Sciences, Universiti Teknologi MARA (Perak), 35400 Tapah Road, Perak, Malaysia*

[3]*Faculty of Applied Sciences, Universiti Teknologi MARA (Perlis), 02600 Arau, Perlis, Malaysia*

Corresponding author. E-mail: hasnunismail@gmail.com

CONTENTS

ABSTRACT

The impact of global warming on aquatic biodiversity is undeniable nowadays. This study emphasized on the effect of temperature to the life performance of tropical cladoceran, *Ceriodaphnia cornuta*. The life performance including longevity, age at first reproduction (AFR), egg development time (EDT), total egg clutch, total number of eggs, and total number of offspring were assessed at selected temperatures (22°C, 26°C, and 30°C). Results showed that the longevity was inversely proportional to an increase in temperature. Meanwhile, AFR and EDT developed faster as the temperature kept increasing. However, the total number of egg clutch, total number of eggs, and total number of offspring showed a unimodal pattern with the peak for reproductive performance at 26°C. Conclusively, temperature has a significant effect on the life performance of *C. cornuta* and this finding contributes to our understanding on consequences of global warming toward aquatic biodiversity.

3.1 INTRODUCTION

Zooplankton are planktonic drifting animals found in many aquatic ecosystems including salt, brackish, and freshwater [1, 4]. They are sufficiently large in numbers and comprise various species of animals mainly dominated by cladocerans, copepods, and rotifers. Zooplankton constitutes an important link in food process in two respects either as grazers (primary and secondary consumers) or serves as food for higher level consumers (planktivores and piscivores).

They are highly responsive to several biotic and abiotic factors such as temperature [2–4], nutrient level [5, 6], light [7], pollution [8], pH level, heavy metal [9], predation [10], and community interactions [11].

Of all the aforementioned factors, temperature is considered to be the most critical factor. Temperature has been shown to affect survival [12], growth [13], metabolism, morphology, reproduction, and behavior in aquatic poikilotherms [14]. Increase in temperature always resulted in the increment of metabolic and growth rates but a decrease in the size of the zooplankton. Conversely, when an upper extreme temperature is reached, the metabolic rates reverse. Nevertheless, the degree to which these biological processes are affected is dependent upon several factors.

Thus, understanding on the thermal regulation in the life cycle of zooplankton is of great importance to reveal their adaptive responses in concern to global warming [15–18].

This investigation was undertaken to quantitatively assess the impact of temperature on the survival, development, and reproduction of freshwater cladoceran, *Ceriodaphnia cornuta*. This species belongs to the same family with daphnids, that is, Daphnidae. We aim to examine *C. cornuta* because it is a dominant zooplankton species naturally distributed in Tasik Elham, Perlis. The result would be useful to provide a fundamental knowledge on the population dynamic and shed light on the use of local species as live-feed source in aquaculture.

3.2 MATERIALS AND METHODS

Female cladocerans were captured from Tasik Elham, Perlis under the water temperature of 26°C. Generations of cladoceran were cultured in batch culture under static condition using aged tap water as the culture medium. The laboratory population was established as a stock culture in a 10 L glass container. Single-celled green algae, *Chlorella* sp. was given as food to the cladocerans at the density of 10^5–10^6 cells/mL. Food regime was offered for every 2 days. The food density is considered as the optimal concentration for all kind of daphnids species [19].

3.2.1 LIFE HISTORY STUDY

Three different temperatures were selected to determine the possible adaptive responses developed in the life cycle of cladoceran (22°C, 26°C, and 30°C). Prior to experimentation, a single female was isolated and adapted to the selected temperature until the third generation was born. Only individuals from the third generation were applicable for life cycle experimentation. This is a standard procedure to offset phenotypic plasticity and maternal effect [20].

Twenty individuals of *C. cornuta* (<24 hours old) were inoculated individually in culture vessel containing 5 mL media. The media was a combination of distilled water and cell suspension of algae offered as food at the optimal density (10^5–10^6 cells/mL). All vessels were incubated in the GC-500 growth chamber within the desired temperatures. Photoperiodsm was set at 12h:12h light and dark cycle with 6000 LUX lighting condition. The medium was changed daily to ensure the animals were free from food limitation and accumulation of waste products.

On daily basis, life cycle parameters were recorded including the age at first reproduction (AFR), egg development time (EDT), total egg clutch (EC), total offspring, and longevity. Newly born neonates and dead individuals were recorded and removed immediately. Recording and observation were conducted using a digital stereomicroscope with camera (SZ51, Olympus) with the highest total magnification of 84×. The experiments were ended when all experimental cladocerans died.

3.2.2 STATISTICAL ANALYSIS

All data were analyzed using the one-way analysis of variance (ANOVA) followed by the post hoc test (Tukey multiple comparison test). The data have been explored and has met the normality assumption for parametric tests. The significant level was set up at $p < 0.05$.

3.3 RESULTS AND DISCUSSION

This study reveals the impact of temperature on most life cycle parameters of *C. cornuta*. This aspect of temperature effect on survival has received a great amount of investigative research. The results revealed that temperature significantly affected all the life cycle parameters of *C. cornuta* (Table 3.1). The life cycle parameters cover the survivorship, development, and reproduction of *C. cornuta*. The longevity represents the survival of *C. cornuta*. The development of *C. cornuta* is shown by the egg development time and age at first maturation. Meanwhile, their reproductive performance is measured from total clutch, total egg, and total offspring.

Data represent degree of freedom, sum of squares, mean square, variance, and level of significant differences ($n = 60$ animals in each group).

TABLE 3.1   Results of ANOVA for Life Cycle Parameters of *C. cornuta.*

One-Way ANOVA	df	SS	MS	F	P
Longevity (d)					
Temperature treatment	2	623.033	311.517	188.999	*0.000
Error	57	93.950	1.648		
AFR (d)					
Temperature treatment	2	7.012	3.506	33.062	*0.000
Error	57	6.045	0.106		
EDT (d)					
Temperature treatment	2	0.067	0.033	19.963	*0.000
Error	57	0.095	0.002		
Total egg clutch (no. female^{-1})					
Temperature treatment	2	26.433	13.217	10.160	*0.000
Error	57	74.150	1.301		
Total egg (no. female^{-1})					
Temperature treatment	2	2309.233	1154.617	25.593	*0.000
Error	57	2571.500	45.114		
Total offspring (no. female^{-1})					
Temperature treatment	2	2310.100	1155.050	28.168	*0.000
Error	57	2337.300	41.005		

Data represent degree of freedom, sum of squares, mean square, variance, and level of significant differences ($n = 60$ animals in each groups).

3.3.1 LONGEVITY

The mean longevity of individuals showed a significant difference among 22°C, 26°C, and 30°C (ANOVA; $p < 0.05$; Table 3.1). The longest longevity occurred at 22°C (18.00 ± 0.38), followed by 11.75 ± 0.28 and 10.75 ± 0.14 at 26°C and 30°C, respectively. The longevity of *C. cornuta* was significantly decreased from 22°C to 30°C (Tukey HSD; $p < 0.05$; Fig. 3.1).

Temperature strongly affects longevity of *C. cornuta*. The longevity was the longest at 22°C and significantly decreased as the temperature increased up to 26°C and 30°C. This finding was in agreement with other tropical cladoceran namely *Ceriodaphnia rigaudi* [3], *Daphnia similis*, *Simocephalus vetulus*, *Moina macrocopa*, and *C. cornuta* [21] that showed similar patterns of longevity. The most optimal temperatures for *D. similis* and *M. macrocopa* were from 18°C to 20°C while 10°C to 12°C for both

S. vetulus and *C. cornuta*. In addition, Nandini and Sarma [22] and Melão and Rocha [23] also observed the same survival patterns for *C. cornuta, M. macrocopa, Pleuroxus aduncus, S. vetulus,* and *Bosminopsis deitersi* in their studies. Similarly, Xi et al. reported significant reductions of *M. macrocopa* life span with increasing of temperature from 18°C to 33°C [24]. This finding suggests that the most optimal temperature for survival of tropical cladocerans lies between 20°C and 30°C. Within suboptimal temperature, the survival of cladocerans is adapted with reduced longevity.

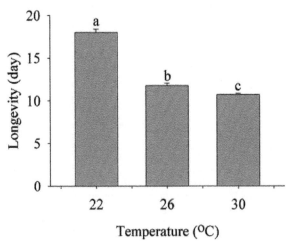

FIGURE 3.1   Responses of life history parameter, longevity (days) to temperature changes.

3.3.2   AGE AT FIRST REPRODUCTION (AFR)

The age at first reproduction (AFR) decreased with increasing temperature from 2.5 days (2.54 ± 0.10) at 22°C to 1.9 days (1.87 ± 0.05) at 26°C followed by 1.8 days (1.77 ± 0.06) at 30°C (ANOVA; $p < 0.05$). The AFR was significantly different between 22°C and 26°C (Tukey HSD; $p < 0.05$), while between 26°C and 30°C the AFR was not significantly different (Tukey HSD; $p > 0.05$; Fig. 3.2).

The results showed that *C. cornuta* matured earlier with increasing temperature. This is more related to the physiological effect of temperature on cladocerans where high temperature accelerates metabolic rates and faster eggs maturation [2]. On the contrary, at low temperature, sexual maturation processes of *C. rigaudi*, *M. micrura*, *Daphnia ambigua,* and *Bosminopsis longirostri* are delayed with longer AFR [3, 25].

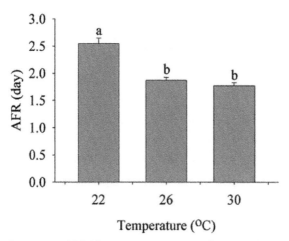

FIGURE 3.2   Responses of life history parameter, age at first reproduction (AFR) (days) to temperature changes.

3.3.3   EGG DEVELOPMENT TIME (EDT)

The results showed that the EDT was significantly shorter between test organisms at 22°C and 26°C (ANOVA; $p < 0.05$). EDT was significantly shorter at 22°C and 30°C too (ANOVA; $p < 0.05$). However, it was not significantly differed between 26°C and 30°C (Tukey HSD; $p > 0.05$).

Furthermore, many studies have shown that the egg development of cladocerans depended on temperature [23, 26]. The EDT (Fig. 3.3) of *C. cornuta* followed similar pattern to their AFR. Our findings revealed that the EDT was inversely related with increasing temperature indicating that eggs develop faster at high temperature. Similarly, the same trend of egg development was found in other tropical cladocerans such as *S. vetulus, M. macrocopa* [21], and *B. deitersi* [23].

Meanwhile, Botrell et al. [27] have reviewed the egg development rates for cladocerans and copepods which developed faster at higher temperature (26°C). In addition, Herzig [28] and Yufera [29] also agreed that eggs development time decreased with increasing temperatures in rotifers. Therefore, it can be inferred that inverse relationship between temperature and the development of cladocerans and the results on AFR and EDT obtained were comparable to other zooplankton species found in literatures.

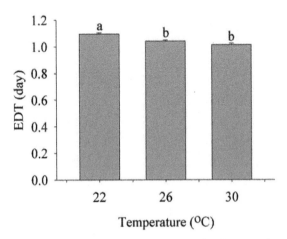

FIGURE 3.3 Responses of life history parameter, egg development time (EDT), (days) to temperature changes.

Even though several researchers suggested the development in freshwater cladocerans itself is affected only by temperature [23, 31, 32], other factor such as food density [22] and predation impact [33, 34, 35] should be taken into consideration. Several tropical cladocerans species showed some interesting results indicating that low food did affect their EDT which might be extended under poor nutritional situation [22].

3.3.4 CLUTCH NUMBER AND EGG PRODUCTION

Mean number of EC showed a significant difference between 22°C and 26°C (ANOVA; $p < 0.05$). Conversely, it does not differ significantly between 26°C and 30°C. Similarly, it was not significantly different between 22°C and 30°C (Tukey HSD; $p > 0.05$). On average, females can produce the highest number of eggs at 26°C with 35.00 ± 1.40 eggs followed by 28.45 ±1.70 eggs and 19.85 ±1.39 eggs at 30°C and 22°C, respectively (Fig. 3.4). The mean number of eggs produced per female also differed significantly among test temperatures (ANOVA; $p < 0.05$; Fig. 3.5).

The clutch number and offspring produced showed a unimodal pattern which increased from 22°C to the highest peak at 26°C and later decreased at higher temperature of 30°C. This pattern was also observed in *D. similis, S. vetulus,* and *M. macrocopa* [21]. Smaller clutches were produced at higher temperatures by other tropical cladoceran too [3].

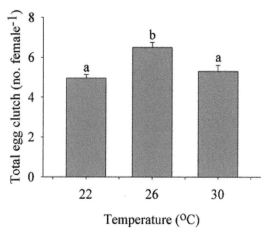

FIGURE 3.4   Responses of life history parameter, total egg clutch number (no. female^{-1}) to temperature changes.

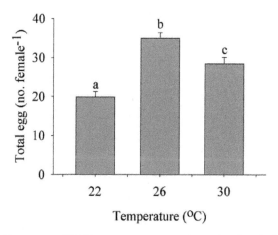

FIGURE 3.5   Responses of life history parameter, total eggs (no. female^{-1}) to temperature changes.

In this study, the unimodal pattern was also observed for the total egg production of *C. cornuta*. *C. cornuta* produced a maximum of 35 eggs at 26°C during its average 12 days lifespan. However, Malhotra and Langer [21] documented that during its average lifespan of 16 days, *C. cornuta* produced the highest number of 47 eggs at higher temperature (30–32°C). In other study on the same species, the number of eggs produced (female^{-1}) were gradually increased, after which experiencing a steady fall (eggs production) till death [30].

3.3.5 OFFSPRING PRODUCTION

Temperature had an impact on offspring production. The mean number of offspring produced per female differed significantly among test temperatures (ANOVA; $p < 0.05$). As the temperature increased from 22°C to 30°C, it significantly decreased the offspring production. The highest number of offspring produced was found at 26°C (Fig. 3.6). On the whole, lowest fecundity was observed in *C. cornuta* at higher temperature. Yet, other than temperature, fecundity of this species was found to be significantly affected by the food level as well [2].

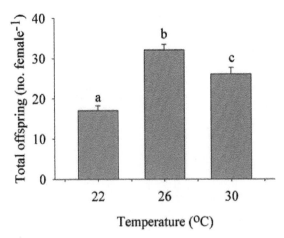

FIGURE 3.6 Responses of life history parameter, total offspring (no. female^{-1}) to temperature changes.

3.4 CONCLUSION

In conclusion, temperature changes significantly affected the life cycle parameters of *C. cornuta*. High temperature accelerates egg hatching, age at maturity, and rate of egg production but shortens lifespan. Meanwhile, lower temperatures usually have the converse effect on the parameters [36]. Consequently, the optimal survival temperature ranges from 22°C to 30°C and optimal life condition was found at 26°C.

3.5 ACKNOWLEDGMENT

This research was funded by UiTM (Perlis) under The Academic Excellent Fund with the project number 600_UiTMPs/PJIM&A/ST/DKCP5/3(06/2012).

KEYWORDS

- *Ceriodaphnia cornuta*
- longevity
- age at first reproduction
- egg development time
- zooplankton

REFERENCES

1. White, P. A.; Kalff, J.; Rasmussen, J. B.; Gasol, J. M. The Effect of Temperature and Algal Biomass on Bacterial Production and Specific Growth Rate in Freshwater and Marine Habitats. *Microbial Ecol.* **1991**, *21*(1), 99–118.
2. Amarasinghe, P. B.; Boersma, M.; Vijverberg, J. The Effect of Temperature, and Food Quantity and Quality on the Growth and Development Rates in Laboratory-cultured Copepods and Cladocerans from a Sri Lankan Reservoir. *Hydrobiologia* **1997**, *350*(1–3), 131–144.
3. Jeronimo, F. M.; Lopez, C. V. Population Dynamics of the Tropical Cladoceran *Ceriodaphnia rigaudi* Richard,1894 (Crustacea: Anomopoda). Effect of Food Type and Temperature. *J. Environ. Biol.* **2011**, *32*, 513–521.

4. Ismail, H. N.; Qin, J. G; Seuront, L. Regulation of Life History in the Brackish Cladoceran, *Daphniopsis australis* (Sergeev and Williams, 1985) by Temperature and Salinity. *J. Plankton Res.* **2011**, *33*(5), 763–777.

5. Jafari, N.; Nabavi S. M.; Akhavan, M. Ecological Investigation of Zooplankton Abundance in the River Haraz, Northeast Iran: Impact of Environmental Variables. *Arch. Biol. Sci. Belgrade.* **2011**, *63*(3), 785–798.

6. Sommer, U.; Adrian, R.; De Senerpont, D. L.; Elser, J. J.; Gaedke, U.; Ibelings, B.; Winder, M. Beyond the Plankton Ecology Group (PEG) Model: Mechanisms Driving Plankton Succession. *Annu. Rev. Ecol. Evol. Syst.* **2012**, *43*, 429–448.

7. Kleiven, O. T.; Larsson, P.; Hobk, A. Sexual Reproduction in *Daphnia magna* Requires Three Stimuli. *Oikos* **1992**, *65*, 197–206.

8. Sakamoto, M.; Tanaka, Y. Different Tolerance of Zooplankton Communities to Insecticide Application Depending on the Species Composition. *J. Eco. Env.* **2013**, *36*(2), 141.

9. Gagneten, A. M.; Paggi, J. C. Effects of Heavy Metal Contamination (Cr, Cu, Pb, Cd) and Eutrophication on Zooplankton in the Lower Basin of the Salado River (Argentina). *J. Water Air Soil Pollut.* **2009**, *198*(1), 317–334.

10. Serpe, F. R.; Larrazábal, M. E. L.; Santos, P. J. P. Effects of a Vertebrate Predator (*Poecillia reticulata*) Presence on *Ceriodaphnia cornuta* (Cladocera: Crustacea) in Laboratory Condition. *Acta Limnol. Bras.* **2009**, *21*(4), 399–408.

11. Mustapha, M. K. Zooplankton Assemblage of Oyun Reservoir, Offa, Nigeria. *Rev. Biol. Trop.* **2009**, *57*(4), 1027–1047.

12. Moore, M. V.; Folt, C. L.; Stemberger, R. S. Consequences of Elevated Temperature for Zooplankton Assemblages in Temperate Lakes. *Arch. Hydrobiol.* **1996**, *135*, 289–319.

13. Hopcroft, R. R.; Roff, J. C.; Bouman, H. A. Zooplankton Growth Rates: The Larvaceans Appendicularia, Fritillaria and Oikopleura in Tropical Waters. *J. Plankton Res.* **1998**; *20*, 539–555.

14. Heinle, D. R. Temperature and Zooplankton. *Chesapeake Sci.* **1969**, *10*(3–4), 186–209.

15. De Stasio, B. T.; Hill, D. K.; Kleinhans, J. M.; Nibbelink, N. P.; Magnuson, J. J. Potential Effects of Global Climate Change on Small North-temperate Lakes: Physics, Fish, and Plankton. *Limnol. Oceanogr.* **1996**, *41*, 1136–1149.

16. Fee, E. J.; Hecky, R. E.; Kasian, S. E. M.; Cruikshank, D. R. Effects of Lake Size, Water Clarity, and Climatic Variability on Mixing Depths in Canadian Shield Lakes. *Limnol. Oceanogr.* **1996**, *41*, 912–920.

17. Stemberger, R. S.; Herlihy, A. T.; Kugler, D. L.; Paulsen, S. G. Climatic Forcing on Zooplankton Richness in Lakes of the Northeastern United States. *Limnol. Oceanogr.* **1996**, *41*, 1093–1101.

18. Chen, C. Y.; Folt, C. L. Consequences of Fall Warming for Zooplankton Overwintering Success. *Limnol. Oceanogr.* **1996**, *41*, 1077–1086.

19. Delbare, D.; Dhert, P. Cladocerans, Nematodesand Trochophora Larvae. In *Manual on the Production and Use of Live Food for Agriculture;* Lavens, P., Sorgeloos, P., Eds.; FAO Fisheries Technical Paper: Ghent, Belgium, 1996; Vol. 361, p. 286.

20. Lynch, M.; Ennis, R. Resource Availability, Maternal Effects, and Longevity. *Exp. Gerontol.* **1983**, *18*, 147–165.

21. Malhotra, Y. R.; Langer, S. Effect of Temperature on Growth and Fecundity of Selected Species of Cladocera. *J. Indian Inst. Sci.* **1993**, *73*, 335–345.

22. Nandini, S.; Sarma, S. S. S. Lifetable Demography of Four Cladoceran Species in Relation to Algal Food (*Chlorella vulgaris*) Density. *Hydrobiologia* **2000**, *435*, 117–126.

23. Melão, M. G. G.; Rocha, O. Life History, Population Dynamics, Standing Biomass and Production of *Bosminopsis deitersi* (Cladocera) in a Shallow Tropical Reservoir. *Acta Limnol. Bras.* **2006**, *18*(4), 433–450.

24. Xi, Y. L.; Hagiwara, A.; Sakakura, Y. Combined Effects of Food Level and Temperature on Life Table Demography of *Moina macrocopa* Straus (Cladocera). *Int. Rev. Hydrobiol.* **2005**, *90*, 546–554.

25. Fileto, C.; Arcifa, M. S.; Henry, R.; Ferreira, R. A. R. Effects of Temperature, Sestonic Algae Features, and Seston Mineral Content on Cladocerans of a Tropical Lake. *Ann. Limnol. Int. J. Lim.* **2010**, *46*, 135–147.

26. Chen, C. Y.; Folt, C. L. Ecophysiological Responses to Warming Events by Two Sympatric Zooplankton Species. *J. Plankton Res.* **2011**, *24*(6), 579–589.

27. Botrell, H. H.; Duncan, A.; Gliwicz, Z. M.; Grygierek, E.; Herzig, A.; Hillbricht-Ilkoska, A.; Kurazawa, H.; Larsson, P.; Weglenska, T. A Review of Some Problems in Zooplankton Production Studies. *Norw. J. Zool.* **1976**, *24*, 12–456.

28. Herzig, A. Comparative Studies on the Relationship Between Temperature and Duration of Embryonic Development of Rotifers. *Hydrobiologia* **1983**, *104*(1), 237–246.

29. Yufera, M. Effect of Algal Diet and Temperature on the Embryonic Development Time of the Rotifer *Brachionus plicatilis* in Culture. *Hydrobiologia* **1987**, *147*(1), 319–322.

30. Subash Babu, K. K.; Nayar, C. K. G. Observations on the Life Cycle of *Ceriodaphnia cornuta* Sars. *J. Zool. Soc. Kerala* **1993**, *3*(2), 13–17.

31. Hall, D. J. An Experimental Approach to the Dynamics of a Natural Population of *Daphnia galeata* Mendotae. *Ecology* **1964**, *45*, 94–112.

32. Achenbach, L.; Lampert, W. Effects of Elevated Temperatures on Threshold Food Concentrations and Possible Competitive Abilities of Differently Sized Cladoceran Species. *Oikos* **1997**, *79*, 469–476.

33. Chang, K. H.; Nagata, T.; Hanazato, T. Direct and Indirect Impacts of Predation by Fish on the Zooplankton Community: An Experimental Analysis Using Tanks. *Limnology* **2004**, *5*, 121–124.

34. Chang, K. H.; Hanazato, T. Prey Handling Time and Ingestion Probability for *Mesocyclops* sp. Predation on Small Cladoceran Species *Bosmina longirostris*, *Bosminopsis deitersi*, and *Scapholeberis Mucronata*. *Limnology* **2005**, *6*, 39–44.

35. Mandima, J. J. The Food and Feeding Behavior of *Limnothrissa miodon* (Boulenger, 1906) in Lake Kariba, Zimbabwe. *Hydrobiologia* **1999**, *407*, 175–182.

36. Sarma, S. S. S; Rao, T. R. Population Dynamics of *Brachionus patulus* Muller (Rotifera) in Relation to Food and Temperature. *Proc.: Animal Sci.* **1990**, *99*(4), 335–343.

CHAPTER 4

SEMANTIC MULTI-MODALITY ONTOLOGY IMAGE RETRIEVAL WITH RELEVANCE FEEDBACK FOR HERBAL MEDICINAL PLANTS

MOHD SUFFIAN SULAIMAN,* SHARIFALILLAH NORDIN, and NURSURIATI JAMIL

Faculty of Computer and Mathematical Sciences, Universiti Teknologi MARA, 40450 Shah Alam, Selangor, Malaysia

Corresponding author. E-mail: suffian@tmsk.uitm.edu.my

CONTENTS

ABSTRACT

With the increasing demand of medicinal herbal plant globally every year, it is vital to preserve it using information technology so that the herbal medicinal plant community as well as next generation can certainly access the information efficiently. Instead of the common textual information, the medicinal herbal plant image is one of the main types of information that people always looking for. It is because human beings acquire the majority of information from the real world from their visual sense. However, to retrieve the required image for an ordinary user is a challenging task. User has a tendency to use high-level feature to interpret the images and measure the similarity rather than using low-level feature. In other words, humans understand the image qualitatively but machine recognizes the image quantitatively. Therefore, we proposed to use semantic multimodality ontology with relevance feedback approach to improve the retrieval efficiency and subsequently measure the effectiveness of retrieval such as precision and recall.

4.1 INTRODUCTION

Image retrieval (IMR) is the field of study concerned with searching, browsing, and retrieving digital images from large database of digital images and the World Wide Web [1]. This is due to the existence of immense Internet technology and advanced electronic image capturing devices such as digital camera, more images have been produced in digital form, daily. Therefore, effective, fast, relevant, and accurate access of digital images become the main goal of IMR from various domains such as remote sensing, digital library, medicine, crime prevention, publishing, architecture, astronomy, agriculture, etc. During the initial stage of IMR, the researcher mostly emphasizes on text-based image retrieval (TBIR). In TBIR, an image uses associated text to determine what the image contains. However, there are several disadvantages using TBIR such as highly labor-intensive and polysemy problem. Therefore the content-based image retrieval (CBIR) research is proposed and many advanced algorithms have been developed and tested to describe the color, shape, and texture features. Currently, IMR research is moving to the semantic-based image retrieval (SBIR) [2] since the CBIR failed to resolve the semantic issues between human and machine. Naturally, humans described an image narratively using high-level semantic visual information, different from machine which is prone to using numeric value as low-level feature. Thus, the semantic gap exists [3]. Semantic gap is "the lack of coincidence between the information that one can extract from visual data and the interpretation that the same data have for a user in a given situation." In other words, the way human interprets the image and how machine extract the image is different. Thus, the SBIR approaches became recognized a few years ago and soon became a prominent research for image and other multimedia information retrieval.

A lot of works have been done in SBIR previously. However, the works are limited to a few domains such as natural scenes [4], sport news [5], animals [6], and furniture catalogs [7]. There is lack of work focusing on herbal medicinal plant so far. Ironically, almost 80% of developed countries depend on traditional medicine for primary health care [8]. The herbal medicines which can be classified as herbs, herbal materials, herbal preparations, and finished herbal products such as food supplements, tonics, nutraceuticals, cosmetics, and toiletries are used to sustain health, prevent, diagnose, improve, or treat physical and mental illnesses. The trends show that the demand will become higher in future and will attract a huge number of people around the world. Thus, it becomes significant and popular domain, serving the interests from huge diversity of herbs community and the necessity to preserve the herbal medicinal information using semantic technology were undeniably.

There are five main categories of SBIR techniques that can be used to reduce the semantic gap [1]:

1) To employ the ontology to define the high-level concepts.
2) To use machine learning techniques to associate low-level features with query concepts.
3) To use relevance feedback (RF) into the system in order to account for the user's actions.
4) To use semantic templates to assist high-level IMR.
5) To use both textual and visual content from the web to assists IMR.

Since we proposed to use multimodality ontology incorporate with RF by exploiting the integration of high-level textual information with low-level visual content, therefore our proposed approach is closest to solution 1 combining with solutions 3 and 5. This combination solution approaches proposal aims to reduce the problem of semantic gap. To the best of our knowledge, there exists no model using the proposed method for herbal medicinal plant domain especially on image dataset.

The rest of this chapter is organized as follows: Section 2 introduces some related works. Section 3 focuses on the discussion of the proposed model, architecture, and methodology. The conclusion and future work are given in Section 4.

4.2   BACKGROUND AND RELATED RESEARCH

4.2.1   ONTOLOGY

Ontology can be defined as a formal depiction of a set of entities within a domain and the relationships among those entities. A formal ontology comprises a controlled vocabulary articulated in a representation language. This language has syntax for using vocabulary terms to label something meaningful within the interest-specified domain [9]. Ontology also used to support the sharing and reuse of formally represented knowledge among artificial intelligent (AI) systems and it can be used to define the mutual vocabulary in which shared knowledge is embodied [10]. Another widely accepted definition, ontology is a specification of a representational vocabulary for a shared domain of discourse which contains the classes, relations, functions, and other objects [10]. In the context of IMR research, ontology was used to define the high-level concepts, so that the machines are readable and processable to understand how humans interpret the images semantically [1, 11, 12]. For example, in some IMR system, image descriptors are used to form a simple vocabulary such as light green, medium green, and dark green [13]. It provides a qualitative definition of high-level query concepts which are understood by humans. Reference [14] implements the ontology in IMR to show how ontologies can help the user in formulating the information need, the query, and the answer. Therefore the ontology can assist the machines to analyze semantic visual information from various sides and give the images unlimited descriptive power of semantics.

4.2.2   MULTIMODALITY WITH ONTOLOGY

The concept of linking high-level textual information and low-level visual feature in IMR was introduced in early 1990s. A pioneering work was published by Pentland in 1995. This concept in Photobook [15] stated that the image content description can be combined with text description to provide a sophisticated browsing and retrieval capability. VisualSEEk [16] also utilizes the concept of combining the text and image features in their search engine. The system tried to match the query-by-example (QBE) with images inside the database based on the low-level feature and the spatial relationship of image region specified by the end user. Some earlier CBIR systems proposed the concept of integrating high-level textual information and low-level feature without considering the need of ontology which has the capability to provide an explicit domain-oriented semantics in terms of defining concepts and their relationship which not only are machine-readable but also machine-processable. Therefore, a multimodality ontology approach was proposed by Wang [17] in 2006 to enhance the previous studies in order to achieve the main goal in IMR. This concept was then exploited by several researchers in their respective IMR system. Multi-modality ontology is the integration of high-level textual information and low-level image feature metadata to represent image contents for IMR [18]. A number of studies [5, 17–21] proved that multimodality ontology can improve the retrieval performance to get more accurate results. Previously, researchers prone to adapt the single ontology approaches, which is textual description ontology. Thus, deny the need of low-level feature. However, [17, 18] performed the experiment by comparing the traditional keyword-based, single text ontology and multimodality ontology. By using the 4000 canine domain images as a sample, he found that the retrieval performances are improved about 5–30% by combining the high-level textual information with low-level image feature and introducing the domain ontology into the multimodality ontology as an important cue to solve the problem of semantic interpretation in IMR. Therefore, the multimodality ontology can provide better retrieval result compared to single ontology. [5, 19] proposed the improvement by integrating the multimodality ontology with DBpedia. The proposed method improved the performance of IMR by interlinking with the Linked Data technologies. Web developers will make use of the rich source of information and the domain ontology to enrich their vocabulary control. [20, 21] utilize the multimodality ontology concept for development of IMR system for their respective images on medicinal dataset. Despite the fact that employing the multimodality ontology in IMR showed a better result compared with the keyword-based and single ontology, there is still a gap to achieve the optimum result. Therefore, this chapter discusses the advantages of multimodality ontology approaches for herbal medicinal plant dataset by combining with RF method.

4.2.3 RELEVANCE FEEDBACK

RF is an interactive mechanism which involved humans as a part of the retrieval process [22]. Su et al. [23] define the RF as a set of methodologies that learn from a collection of users' browsing behaviors on IMR. In such CBIR system that uses RF, the scenario can be depicted as a user marked the queried images as a searched image that they wanted. The images then fed back into the systems as a new redefined query for the following cycle of retrieval process. This operation is iterated until the user is satisfied with the end query result [22]. Therefore, RF has the capability to assist the user in IMR to get the smallest possibilities queried results as what user looking for and increased the quantitative measurement, that is, precision of IMR.

There are six RF techniques commonly used in CBIR [22]:

1) Query reweighting (QR).
2) Query point movement (QPM).
3) Query expansion (QE).
4) Log-based RF.
5) Navigation pattern relevance feedback (NPRF).
6) Particle swarm optimization RF.

The previous multimodality ontology model discussed dealt with either improving the content of multimodality ontology or utilizing it in different domains. The researchers contributed a great attention to enrich the vocabulary of ontology to resolve the lack of depth metadata in ontology with the hope that the end user would retrieve the image successfully. However, lack of work has been done to emphasize on the accuracy of retrieval mechanism especially when the end user confront with a large scale of image metadata. Despite the end user still able to retrieve the image as they wanted, the number of potential retrieved image would vary especially when the result of queried images is enormous. Since the RF has the capability to maximize the possibility for end user to obtain the targeted image precisely, there is a need to integrate the multimodality ontology with RF. So, we propose to incorporate the multimodality ontology with NPRF. We choose NPRF since it is proven that it is better than other techniques in terms of precision, coverage, and the number of feedbacks [22].

4.3 PROPOSED MODEL

4.3.1 THE MODEL

Figure 4.1 depicted the main components of our proposed conceptual framework. The architecture of the framework is divided into three main components:

1) Knowledge extraction.
2) Knowledge base construction.
3) Retrieval mechanism with RF.

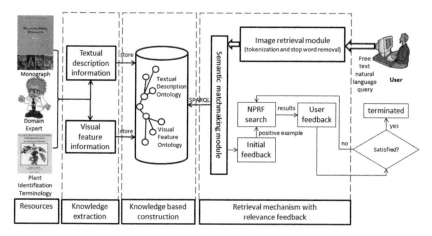

FIGURE 4.1 Architecture of multimodality ontology with relevance feedback.

(1) Knowledge extraction

The knowledge extraction component is divided by analyzing of textual description information and interpretation of the semantic visual feature information. The analysis of textual description information begins with collecting the information from corpora as a main resource such as Malaysian Herbal Monograph volumes I, II, and III [24] as a dataset. The scope is focused on the knowledge of plant description, external morphology of herbal plant, and traditional uses. More than 60 herbal medicinal plant species to be analyzed comprise leaf, rhizome, fruit, root, and flower. We also refer to the plant identification terminology [25] and domain experts, that is, research officer from Medicinal Plants Division, Forest Research Institute Malaysia (FRIM) because the monographs contained the domain-oriented information. Based on the collected information, we analyze and extract the knowledge before constructing the ontology. This task is very crucial because if we misunderstood the botanical term we will produce incorrect ontology. Therefore we need to work closely with the domain expert. Several elicitation techniques such as formal interviews, brainstorming, formal and informal analysis of texts undergo with the domain experts. In this module, we identify the knowledge-based conceptualization, that is, terminology (TBox) by organizing and transforming an informally perceived view of a domain into a semiformal specification. It can be done using a set of intermediate representation which both ontologist and domain experts can understand. Table 4.1 depicted the output of the knowledge extraction for textual description information.

TABLE 4.1 Example TBox Classes of Textual Description Information.

Ontology	Superclass	Subclass	Subclass
Textual concept	Plant	Herb	
		Plant description	General habitat
			Leaf
			Fruit
			Root
			Flower
			Stem
		Synonym	
		Vernacular name	Malay
			English
		Traditional uses	
		Definition	
		Character	Physical
			Odor
			Taste

For the second subcomponent, to interpret the semantic visual feature information, we first define the set of terms adapted from monographs, plant identification terminology, and domain experts which are relevant to the image content. We assume these terms can be distinguished using low-level feature, that is, color, shape, and texture [26]. The extraction and classification is done manually and eventually each image has a set of labels to describe its content, which are to be matched with the concepts defined in the visual feature ontology. The example of TBox for visual feature information and sample of visual features are presented in Table 4.2 and Figure 4.2, respectively.

TABLE 4.2   Example of TBox Classes of Visual Feature Information.

Ontology	Superclass	Subclass
Visual concept	Color	Color leaf
		Color fruit
		Color root
		Color flower
		Color stem
	Shape	Shape leaf
		Shape fruit
		Shape root
		Shape flower
		Shape stem
	Texture	Texture leaf
		Texture fruit
		Texture root
		Texture flower
		Texture stem
	Image	

FIGURE 4.2   Sample of herbal medicinal plant images, that is, the visual features.

(2) Knowledge-based construction

In this approach, we convert the low-level features of images to a set of terms and integrate which can be utilized by semantic matchmaking module. Thus, incorporate the high-level textual information with low-level image attributes. Once all the set of intermediate representation, that is, TBox and assertion (ABox) is done, the module of knowledge-based construction will be implemented. The selected ontology engineering tool, that is, protege will be chosen and the semantic textual and visual metadata will be constructed. Then the integration of the knowledge based between textual description and visual features will be implemented by connecting it using properties and relationships to be a multimodality ontology.

(3) Retrieval mechanism with RF

The retrieval mechanism acts like the search engine. The user can freely enter the semantic query, that is, "a grass-like rhizome with bright green leaf," and IMR module will perform the tokenization and remove the stop words, that is, "a" and "with" before proceeding to semantic matchmaking module. The user-friendly interface needs to be developed for this purpose. This module provides the matching approach of the input query with the multimodality ontology which stored as Resource Description Framework (RDF) consists of triplet; subject–predicate–object. For example, the aforementioned descriptive query components are based on the combination of textual concept; grass-like rhizome as general habit subclass and visual concept; bright green as color leaf subclass. The SPARQL query language can be used to express queries across diverse data sources in multimodality ontology. The result of the semantic query through SPARQL is the retrieval of relevant images which called the initial feedback will be displayed to the user. In this RF process, the user will be iteratively queried until satisfied with the image that the user is looking for as illustrated in Figure 4.1.

4.3.2 MEASUREMENT AND BENCHMARK

For the performance evaluation, we will validate the performance of multi-modality ontology semantic retrieval by using qualitative and quantitative measurement. For the qualitative measurement, two types of evaluation need to be done, which are the correctness of the ontology notation and reviewing the content of the ontology with domain expert. For the quantitative measurement, we choose the precision and recall evaluation to be implemented. Finally, we will benchmark the performance of multimodality ontology with RF with the multimodality ontology without using RF to prove the semantic IMR efficiency as a result.

4.4 CONCLUSION AND FUTURE WORK

This chapter proposed an enhanced multimodality ontology model by incorporating with RF. Based on our review, we choose NPRF since it is better in terms of the retrieval efficiency among other techniques. NPRF will be used to integrate with textual description ontology and visual feature ontology. Since the ontology development is uniformly constrained with domain oriented, the herbal medicinal plant domain is choosen for this research. In this chapter, we proposed the architecture design of multimodality ontology with RF, identified the related resources, extracted the textual description, and visual feature information and constructed a part of the knowledge based particularly on the TBox. For the next research phase, we plan to complete the construction of knowledge based particularly on the ABox, construct the retrieval mechanism and RF and further our work with development of the prototype of multimodality ontology with RF and finally measure the performance of this model.

4.5 ACKNOWLEDGMENT

The authors would like to thank Universiti Teknologi MARA, Malaysia and the Ministry of Higher Education Malaysia for providing the research grant (Grant No. RAGS/2012/UiTM/SG05/1).

KEYWORDS

- **multimodality ontology**
- **semantic image retrieval**
- **relevance feedback**

REFERENCES

1. Liu, Y.; Zhang, D.; Lu, G.; Ma, W. Y. A Survey of Content-based Image Retrieval with High-level Semantics. *Pattern Recogn.* **2007,** *40*(1), 262–282.
2. Magesh, N.; Thangaraj, P. Semantic Image Retrieval Based on Ontology and SPARQL Query. *Int. J. Comput. Appl.* **2011,** *1*(3), 12–16.
3. Smeulders, A. W. M.; Worring, M.; Santini, S.; Gupta, A.; Jain, R. Content-based Image Retrieval at the End of the Early Years. *IEEE T. Pattern Anal.* **2000,** *22*(12).
4. Majeed, S.; Qayyum, Z.; Sarwar, S. In *SIREA: Image Retrieval Using Ontology of Qualitative Semantic Image Descriptions,* International Conference on Information Science and Applications, Suwon, South Korea, 2013, pp 1–6.
5. Khalid., Y. I. A. M.; Azman, S.; Noah, M. Improving the Performance of Multi-modality Ontology Image Retrieval System Using DBpedia. *Procedia Inform. Technol. Comput. Sci.* **2012,** 1–9.
6. Gowsikhaa, D.; Abirami, S.; Baskaran, R. In *Construction of Image Ontology Using Low-level Features for Image Retrieval,* International Conference on Computer Communication and Informatics, Coimbatore, India, 2012, pp 1–7.
7. Andreatta, C. In *Ontology Based Viso-semantic Similarity for Image Retrieval,* International Workshop in Content-based Multimedia Indexing, London, UK, 2008.
8. Word Health Organisation (WHO), Traditional Medicine, 2008. [Online]. Available: http://www.who.int/mediacentre/factsheets/fs134/en/# (accessed Sep 22, 2012).
9. Nwogu, I.; Govindaraju, V.; Brown, C. In *Syntactic Image Parsing Using Ontology and Semantic Descriptions,* IEEE Computer Society Conference on Computer Vision and Pattern Recognition Workshops, San Francisco, CA, USA, 2010, pp 41–48.
10. Gruber, T. R. A Translation Approach to Portable Ontology Specifications. *J. Knowl. Acquis.* **1993,** *5*(2), 199–220.
11. Manzoor, U.; Ejaz, N.; Akhtar, N. In *Ontology Based Image Retrieval,* 7th International Conference for Internet Technology and Secured Transactions, London, UK, 2012, pp 288–293.
12. Ye, L.; Hong, B.; Hong-zhe, L. In *The Design and Implementation of Ontology Based Image Retrieval for Xu Beihong Museums,* 2nd IEEE International Conference on Computer Science and Information Technology, Beijing, China, 2009, pp 191–195.
13. Mezaris, V.; Kompatsiaris, I.; Strintzis, M. G. In *An Ontology Approach to Object-Based Image Retrieval,* International Conference on Image Processing, ICIP, 2003, pp 511–514.
14. Hyvonen, E.; Styrman, A.; Saarela, S. Ontology-based Image Retrieval. WWW (Posters), 2003.
15. Pentland, A.; Pizard, R. W.; Sclaroff, S. In *Photobook: Content-based Manipulation of Image Databases,* SPIE Storage and Retrieval Image and Video Databases II, 1995, 255, pp 1–33.
16. Smith J. R.; Chang, S. In *VisualSEEk: A Fully Automated Content-based Image Query System,* Proceedings of the 4th ACM International Conference on Multimedia, Boston, Massachusetts, USA, 1996, pp 87–98.
17. Wang, H.; Liu, S.; Chia, L. In *Does Ontology Help in Image Retrieval ? A Comparison Between Keyword, Text Ontology and Multi-modality Ontology Approaches,*

Proceedings of the 14th annual ACM International Conference on Multimedia, Santa Barbara, CA, USA, 2006, pp 109–112.

18. Wang, H.; Liu, S.; Chia, L. T. Image Retrieval with a Multi-modality Ontology. *J. Multimedia Syst.* **2008**, *13*, 379–390.

19. Kesorn, K. *Multi Modal Multi Semantic Image Retrieval.* University of London, 2010.

20. Singh, P.; Goudar, R.; Rathore, R.; Srivastav, A.; Rao, S. In *Domain Ontology Based Efficient Image Retrieval,* Proceedings of 7th International Conference on Intelligent Systems and Control, Coimbatore, India, 2013.

21. Singh, P.; Rathore, R.; Chauhan, R. Ontology Based Retrieval for Medical Images Using Low Level Feature Extraction. In *Eco-friendly Computing and Communication Systems;* Mathew, J., Patra, P., Pradan, D. K., Kuttyamma, A. J., Eds.; Springer: Berlin, Heidelberg, 2012; pp 413–421.

22. Khalid, Y. I. A. M.; Noah, S. A.; Abdullah, S. N. S. Towards a Multimodality Ontology Image Retrieval. In *Visual Informatics: Sustaining Research and Innovations;* Zaman, H. B., Robinson, P., Petron, M., Oliver, P., Shih, T. K., Velastin, S., Nystrom, I. Eds.; Springer: Berlin, Heidelberg, 2011; Vol. 7067, pp 382–393.

23. Sivakamasundari, G.; Seenivasagam, V. In *Different Relevance Feedback Techniques in CBIR: A Survey and Comparative Study,* International Conference on Computing, Electronics and Electrical Technologies, Kumaracoil, India, 2012, pp. 1115–1121.

24. Su, J.; Huang, W.; Yu, P. S. Efficient Relevance Feedback for Content-Based Image Retrieval by Mining User Navigation Patterns. *IEEE T. Knowl. Data Eng.* **2011**, *23*(3), 360–372.

25. M. H. M. Commitee, Malaysian Herbal Monograph, Volume I. Forest Research Institute Malaysia (FRIM), 1999.

26. M. H. M. Commitee, Malaysian Herbal Monograph, Volume II. Forest Research Institute Malaysia (FRIM), 2009.

27. M. H. M. Commitee, Malaysian Herbal Monograph, Volume III. Forest Research Institute Malaysia (FRIM), 2012.

28. Harris, J. G.; Harris, M. W. *Plant Identification Terminology: An Illustrated Glossary*, Second Ed. Spring Lake Publishing: Spring Lake, Utah, 2004.

29. Sulaiman, M. S.; Nordin, S.; Jamil, N. Plant Image Ontology, In *Computer and Mathematical Sciences Graduates National Colloqium (SiSKOM)*, Shah Alam, Selangor, Malaysia, 2013.

CHAPTER 5

EFFECTS OF DIETS CONTAINING EFFECTIVE MICROORGANISMS (EM) ON GROWTH PERFORMANCE, WORM BURDEN, HEMATOLOGICAL AND BIOCHEMICAL ANALYSIS IN FEMALE GOATS

MUHAMMAD ANBARIQ ABDUL RAZAK[1],
MUHAMMAD ZAKI ZAKARIA[1], MOHD ASMAWI MOHD TAYID[2],
and SHAMSUL BAHRIN GULAM ALI[1,*]

[1]*School of Biology, Faculty of Applied Sciences, Universiti Teknologi MARA, 40450, Shah Alam, Selangor, Malaysia*
[2]*Department of Veterinary Services Kuala Langat, Jalan Sungai Buaya, 42600, Sungai Jarum, Kuala Langat, Selangor, Malaysia*
Corresponding author. E-mail: sbahrin@salam.uitm.edu.my

CONTENTS

ABSTRACT

The aim of this study was to evaluate the effectiveness of diets containing effective microorganisms (EM) in female goats. During five weeks of the experimental period, the weight of goats (kg), fecal egg count (FEC) (epg), anemic status (FAMACHA evaluation), hematological and biochemical parameters were analyzed. The results showed an increase in body weight (kg) and improved anemic status in EM-treated female goats ($p > 0.05$). Moreover, FEC showed a reduction in EM-treated group ($p > 0.05$). Hematological analysis demonstrated an elevation in red blood cell (RBC) and hematocrit (HCT) count but decrease in hemoglobin (Hb) count ($p > 0.05$). White blood cell (WBC) count remained within normal limits for all experimental groups ($p > 0.05$). Biochemical analysis showed decreased blood glucose and cholesterol (total cholesterol, triglycerides, and high-density lipoprotein) levels in EM treated groups. All experimental goats demonstrated normal total protein levels. In conclusion, EM promote fast growth performance, lower the worm burden, improve the anemic status, and improve essential blood metabolites in female goats.

5.1 INTRODUCTION

Livestock production plays an important role in the agricultural and rural economies in the developing world. Livestock not only provides food but also supports the inputs for crop production, agricultural industry, and also employment. This sector employs at least 1.3 billion people worldwide and supports 600 billion poor smallholder farmers in developing world [1]. As a result of increased income and urbanization, a demand-driven Livestock Revolution is underway in developing world with intense implications for agriculture, health, and the environment [2].

The livestock industry in Malaysia is an important subsector in agriculture to be addressed in the Agriculture National Key Economic Areas for food safety and security. Infectious diseases, poor management and nutrition, and inconsistent breeding policies are the main constraints hindering the productivity of small ruminats. Parasitic helminths, coccidian, and blood protozoa are commonly detected in small ruminants, clinically followed by a reduction in immunity leading to bacterial and viral conditions [36].

The total number of sheep and goats in Malaysia has increased from 538,538 heads in 2007 to 562,554 heads in 2008. This number, which constitutes the small ruminant industry of Malaysia, is an important component of the overall livestock sector. One of the main problems limiting productivity in small ruminants of Malaysia is gastrointestinal parasite infections that can cause mortality and morbidity in goats and sheep [3].

Determination of main hematological parameters of animals helps veterinarians to confirm clinical diagnoses, estimate the severity of the disease, administer appropriate treatment, and evaluate health status outcomes [4]. Information gained from blood parameters would substantiate the physical examination and coupled with medical history provides an excellent basis for medical judgement in goats [5]. Biochemical analysis is a test that measures the chemical substances carried by the blood. Some of the important parameters in determining ruminant helminthiasis are glucose, total protein, and blood cholesterol [6].

Effective microorganisms (EM) have a wide field of applications in the environment, animal health, and industrial field. In the agricultural field, the use of EM had improved the quality of soil for better plant growth, treating wastewater, and controlling pest and diseases [7], while in the animal management field, EM increased the quality of meat and production of dairy eggs in poultry. Animals such as goats that grazed on EM-treated pasture and drinking water have a higher liveweight gain compared to the one that grazed on pasture and drinking water without EM [8]. Sheep that was supplemented with EM also shows a reduction of parasites in the body [9]. A preliminary study was conducted on Boer goats supplemented with EM Bokashi had shown a reduction in fecal egg count (FEC), improved FAMACHA scores, and increase in packed cell volume (PCV) percentage [3]. Therefore, the objective of this study was to evaluate the effectiveness of different types of EM (EM Bokashi and EM Activated Solution or EMAS) on the growth performance, worm burden reduction, and hemato-biochemical parameters of female goats.

5.2 MATERIALS AND METHODS

5.2.1 DESCRIPTION OF STUDY SITE

The study was conducted at Baiduri Farm located in Kampung Jawa, Klang from May to June 2013.

5.2.2 EXPERIMENTAL GOATS AND MANAGEMENT

The experimental protocol used in this study was approved by the Department of Veterinary Services of Selangor. Twenty-four mixed breed Boer and Saanen female goats at the age of 1–2 years, weighing between 20–40 kg were randomly allotted into four groups with six goats per group. Group A, was fed once a day with 5% of EM Bokashi mixed with 2.5 kg of commercial pellets and were given normal drinking water. Group B, received normal commercial pellets with drinking water mixed with 20 mL EM EMAS in 20 L of drinking water (1:1000). Group C (positive control) was given anthelmintic drugs and Group D (negative control) received normal food and drinking water. All experimental animals were not allowed to graze freely and were kept in raised floor pens throughout the study. All groups were examined for their growth performance (kg), FEC (egg per gram; epg), FAMACHA anemic evaluation, and hematological and biochemical analysis.

5.2.2.1 BODY WEIGHT (GROWTH PERFORMANCE)

Initial and final individual body weights of all twenty-four female goats were recorded during the study.

5.2.2.2 FECAL EGG COUNT (MODIFIED MCMASTER TECHNIQUE)

Fecal samples were collected in the morning from each goat at week 0, week 1, and week 4. These samples were subjected to modified McMaster technique of fecal egg counting using 2 g of individual fecal samples [3].

5.2.2.3 FAMACHA EVALUATION

The lower eyelids of all animals was pulled down gently with the index finger to expose the goat's ventral conjunctiva and the color of the ocular mucous membrane was compared with a standard FAMACHA chart, and this indicates anemia caused by the helminth infection [3]. The anemic status was evaluated before and after the treatment.

5.2.2.4 HEMATOLOGICAL AND BIOCHEMICAL ANALYSIS

Blood from jugular vein was withdrawn into EDTA blood tubes twice from all goats (pre- and posttreatment) for analysis. Whole blood was subjected to hematological analysis within 24 h after blood sampling. Erythrocytes (RBC), leukocytes (WBC), hematocrit (HCT), and hemoglobin (Hb) counts were analyzed using Boule Medonic Vet Blood Analyzer (CA-530). Remaining blood samples were analyzed for glucose, total protein, total cholesterol, triglycerides (TG), and high-density lipoprotein (HDL) using a clinical chemistry analyzer (Fuji Dri-Chem 400i) according to the manufacturer's protocol and the results were recorded. Quantitative results were expressed as mean ± SEM. All data were analyzed by analysis of variance (ANOVA). Significant differences between control and treatment were determined using Student's t-test with a 5% probability.

5.3 RESULTS AND DISCUSSION

5.3.1 BODY WEIGHT (GROWTH PERFORMANCE)

After 5 weeks of treatment, groups fed with EM (A and B) experienced body weight increment, while body weight reduction ($p > 0.05$) was observed in unfed groups (C and D) (Table 5.1). Previous study on Boer goats [3] and mixed breed Boer goats [10] fed with 5% Bokashi mixed with commercial pellets for 6 weeks showed significant increase in body weight. We believed that duration of EM supplementation may contribute to significant body weight gained in animals. It is possible that EM could have increased the appetite of goats, thus increasing the feed intake and further improving the ruminant digestion process. Another study also stated that food supplemented with EM positively enhanced body weight gain in broilers [11].

Ruminants such as goats require dietary energy source for weight gain [12]. Our study has provided additional evidence that EM can potentially provide high-energy diet which is important in stimulating body weight gain in goats. Moreover, our finding in unfed EM groups supported other finding that claimed the absence of nutritional food resources can seriously reduce the growth performance in goats [13]. It is suggested that EM can be used as feed or water additive for improving body weight performance in animals, which directly provides the dietary energy source.

On the other hand, the naturally occurring microorganisms present in EM after they enter into the body create more effective intestinal microflora with a greater synthetic capability, that is, one that can synthesize hormones, vitamins, and enzyme systems that improve digestion, enhance growth, provide disease resistance, suppress malodor, inhibit pathogen, and improves product quality. Previous studies have shown that supplementing animal diets with EM brought positive results in many animal performances, where an increased rate of egg laying in chickens [7] and a significant individual weight gain in piglets [34], sheep, and goats [35] have been reported.

TABLE 5.1   Body Weight Gain in Goats Fed Diets with EM.

Groups	Week 0 (Pre)	Week 4 (Post)
A	24.17 ± 2.30	26.50 ± 2.81
B	23.17 ± 1.64	24.50 ± 2.14
C	29.67 ± 1.61	27.33 ± 1.50
D	31.50 ± 2.47	30.83 ± 2.65

The values are mean ± SEM. There is no significant difference in weight of goats between weeks and groups ($p > 0.05$).

5.3.2   FECAL EGG COUNTS

The interpretations for FEC analysis for all groups in the present study were considered to have less worm burden (FEC < 500 epg) (Table 5.2) [14]. Based on these results, weekly FEC analysis was not recommended by the local veterinarian [15]. The FECs in Group A dropped steadily from 475 epg (pretreatment) to 300 epg (posttreatment) ($p > 0.05$). A similar finding of an FEC reduction in goats fed with 5% EM Bokashi in food was also observed in the previous study [3, 10]. In Group B animals, increased egg counts (from week 0 to week 1) were probably due to the time needed for EM EMAS to be adapted by the goat's gastrointestinal system.

This finding was similar to the previous study by [16] who suggested that the taste of EM EMAS mixed in water may lower the drinking consumption rate in goats. Some goats in Group B were seen to have less drinking behavior, which could explain the increased of FEC in week 1.

After week 1, the goats may have tolerated with an EM EMAS taste that finally bring to the reduction of FEC from week 1 to week 4.

TABLE 5.2   FEC in Goats Fed Diets with EM.

Groups	Week 0 (Pre)	Week 1	Week 4 (Post)
A	475 ± 46.1	300 ± 54.8	300 ± 77.5
B	125 ± 46.1	275 ± 46.1	175 ± 46.1
C	275 ± 71.6	525 ± 270.4	425 ± 98.1
D	425 ± 90.1	275 ± 60.2	325 ± 46.1

The values are mean ± SEM. There is no significant difference in FEC (epg) between weeks and groups ($p > 0.05$).

Data gathered from the previous study showed that FEC were subjected to a within-individual variation due to factors such as weather, season, random day-to-day variation, and phase of parasitic infections [17]. FEC was also influenced by temperature and moisture, where they were highest in the wet condition. The fact that our research was conducted during the dry season could strongly explain the lower egg counts (FEC < 500 epg) obtained at the beginning of the study. Non EM supplemented groups showed fluctuating egg counts. The output of eggs is also known to fluctuate depending on the physiological status of the hosts [18], and an increase in FEC was previously observed in the late-pregnant and lactating ewes [19]. Lactating activity in some goats in Group D could be attributed to the increase in FEC. During posttreatment, the highest FEC was recorded in Group C (425 epg), treated with an anthelmintic drug, where the count was close to the moderate infection cut-points (FEC: 500–1000 epg) that may indicate resistance development toward anthelmintic drug used [20].

5.3.3   FAMACHA EVALUATION

FAMACHA monitoring was performed before and after the treatment. Four out of six goats in Group A with an increased HCT value showed nonanemic status. In addition, there was a positive correlation in the previous study between improved FAMACHA scores and increased PCV values [3].

During the postanemic evaluation, only one out of six female goats in Group B showed postnonanemic status, five goats were diagnosed to have a postanemic status, in which three goats were mildly anemic and two goats were anemic. These results would raise the possibility that EM EMAS does not rapidly improve the preanemic status, with an addition to the less drinking behavior of goats in group B.

All goats from Groups C (positive control) and D (negative control) showed postmildly anemic and postanemic status. The effectiveness of dewormer is restricted by the frequent use of anthelmintic drug [21]. Frequent usage of anthelmintic drugs, practiced by farm owner was suspected to increase the chances of developing anthelmintic resistance in Group C animals [22].

EM can be used as an alternative to improve the general health and immunity of animals, thereby making them more resilient to parasitic infections [3]. Nevertheless, FAMACHA evaluation scores must not be the sole basis of determining helminth infection in goats. Other signs of gastrointestinal infections such as the FEC and analysis should also be considered when observing animals during handling [3, 14].

5.3.4   HEMATOLOGICAL ANALYSIS

RBC and HCT (or PCV) counts were increased, while Hb counts were decreased ($p > 0.05$) in both supplemented EM groups. However, decreased (RBC, HCT, and Hb) counts were observed in nonsupplemented EM groups ($p > 0.05$). Based on these results, we believed that EM elevates the HCT count, which is an indirect measurement of the iron status balance in goats. Numerous clinical signs are associated with iron deficiencies, including anemia, reduced growth, and increased rate of diseases [23]. Absence of EM may impair iron status balance at the optimum level, where it can be demonstrated in body weight and HCT reduction in nonsupplemented goats.

Increased RBC value for both Groups A and B indirectly indicates that there was a mild infection of gastrointestinal parasite during post-treatment. Our study confirmed the presence of *Haemonchus contortus* (blood-sucking nematode) by the ova identification during FEC analysis. Reduced FEC, improved anemic status in the FAMACHA evaluation with positive correlation to HCT, and increased RBC values were observed in EM-treated groups (Table 5.3).

TABLE 5.3   Hematological Parameters in Goats Fed Diets with EM.

Groups	Hematological Parameters						WBC (106/µL)
	RBC (106/µL)		HCT (%)		HB (g/dL)		Post
	Pre	Post	Pre	Post	Pre	Post	
A	1.14 ± 0.1	1.34 ± 0.1	5.08 ± 0.3	5.87 ± 0.4	8.42 ± 0.3	8.23 ± 0.4	20.22 ± 1.65
B	1.32 ± 0.1	1.48 ± 0.1	5.80 ± 0.6	6.45 ± 0.5	8.75 ± 0.5	8.53 ± 0.4	21.46 ± 1.33
C	1.50 ± 0.2	1.46 ± 0.1	6.55 ± 0.8	6.40 ± 0.6	8.52 ± 0.5	7.98 ± 0.4	20.46 ± 2.19
D	1.19 ± 0.2	1.11 ± 0.2	5.25 ± 0.9	4.90 ± 0.9	7.90 ±0.4	7.27 ± 0.5	16.96 ± 2.50

The values are mean ± SEM. There is no significant difference in hematological parameters between weeks and groups ($p > 0.05$).

Hb values were decreased in all groups of animals. These data suggested that while consuming the host blood, *H. contortus* may destroy Hb indirectly. A previous study has revealed a marked reduction in HCT, Hb, and RBC count in lambs in relations to nematode infections [24]. The reduced HCT, Hb, and RBC counts in infected sheep may be attributed to the bleeding of abomasa due to the injuries caused by this parasite. Under heat stress, undernourished goats are known to have low HCT and Hb values due to compensation to protect internal homeostasis and normal blood levels [25]. Nonsupplemented EM groups were considered to be undernourished due to absence of EM in their diet.

Only post-WBC count data were presented in the study because the animals from the farm were considered to have low worm burden (FEC < 500 epg) that did not reflect any leukocytosis. WBC data gave a good indication of the ability of infected animals to resist infections [26]. WBC levels in all experimental groups were within the normal range, and this may suggest that goats in this study probably had developed a good resistance against gastrointestinal parasite infection. Furthermore, good animal husbandry practice of the study area was well maintained by the owner.

5.3.5   BIOCHEMICAL ANALYSIS

Age, sex, and the act of lactating can affect the pattern of glucose values in goats, where aging and lactating goats showed decreased glucose levels [27, 28, 29]. Some of the goats in Group D were milking the kids, where this could raise the possibility of glucose level reduction (Table 5.4). Lactating goats showed low glucose concentrations due to tremendous

TABLE 5.4 Biochemical Parameters in Goats Fed Diets with EM.

Groups	Biochemical Parameters								
	Glucose (mg/dL)		High-density Lipoprotein (HDL) (mg/dL)		Triglyceride (TG) (mg/dL)		Total Protein (TP) (g/dL)	Total Cholesterol (TC) (mg/dL)	
	Pre	Post	Pre	Post	Pre	Post	Pre and Post	Pre	Post
A	73.3 ± 4.4	28.3 ± 0.7	65.0 ± 6.0	59.7 ± 11.0	31.0 ± 0.0	29.0 ± 4.0	>11	172.5 ± 5.5	143.5 ± 40.5
B	75.0 ± 2.1	25.3 ± 0.9	60.7 ± 2.8	48.0 ± 4.6	41.5 ± 0.5	19.0 ± 2.0	>11	184.5 ± 2.5	143.5 ± 1.5
C	71.7 ± 3.4	23.5 ± 2.5	64.0 ± 6.7	58.7 ± 8.8	39.5 ± 14.5	53.5 ± 1.5	>11	168.5 ± 13.	159.5 ± 15.5
D	62.3 ± 3.3	27.5 ± 1.5	58.3 ± 3.8	53.3 ± 7.1	33.5 ± 10.5	25.5 ± 4.5	>11	179.5 ± 13.	157.5 ± 30.5

The values are mean ± SEM. There is no significant difference in biochemical parameters between weeks and groups ($p > 0.05$)

pressure exerted in order to meet the increased metabolic demand on glucose for lactose synthesis [29]. However, decreased glucose levels were also observed in three other experimental groups (A, B, and C) ($p >$ 0.05). To date, the effect of EM on glucose remains unclear.

Our results demonstrated that both supplemented EM groups were showing decreased levels of TG and HDL (Table 5.4). Probiotic supplementation may depress the concentration of cholesterol due to the incorporation of cholesterol into the cellular membrane of the organism, which in turn reduces the cholesterol absorption in the gastrointestinal tract [30]. Moreover, the yeast, *Saccharomyces cerevisiae* strains was also known to be able to remove cholesterol, and it seems to be a good candidate organism for lowering cholesterol level in the gastrointestinal tract [31]. Hence, it was clearly observed that supplementation of EM in the goats' diets can reduce the level of blood cholesterol.

In total protein analysis, there is no significant difference between pre- and posttreatment in all experimental groups ($p > 0.05$). In addition, the normal level of total protein in goats during the dry season was reported to be 12.0 g/dL [32]. Our study was conducted during the dry season, and the total protein results obtained were within the normal range. Previous study of total proteins on African dwarf goats supplemented with natural legume, *Afzelia africana* and natural shrub, *Newbouldia laevis* did not differ significantly with the control groups [33].

5.4 CONCLUSION

Our experiment reveals that both EM (EM BOKASHI and EMAS) are capable of enhancing goats' body weight, reducing the FEC of helminth, improving FAMACHA scores, and producing better hematological and biochemical profiles. Further study is proposed to the detailed investigation on complete hematobiochemical profiles for a better understanding of the EM mechanism in producing healthy ruminants.

5.5 ACKNOWLEDGMENTS

The author would like to thank Universiti Teknologi MARA, Malaysia for financial assistance and assisting in the publication and the Department of Veterinary Services (DVS) of Selangor for supplying the EM (Bokashi and EMAS); and special thanks to Baiduri Goat Farm for providing an enabling environment for this study.

KEYWORDS

- effective microorganisms Bokashi
- effective microorganisms-activated solution
- helminth
- fecal egg count
- livestock

REFERENCES

1. Thornton, P. K. Livestock Production: Recent Trends, Future Prospects. *Phil. Trans. R. Soc.* **2010**, *365*, 2853–2857.
2. Khan, A. A.; Bidabadi, F. S. Livestock Revolution India: Its Impact and Policy Response. *South Asia Res.* **2004**, *24*(2), 99–122.
3. Chandrawathani, P.; Nurulaini, R.; Premaalatha, B.; Zaini, C. M.; Adnan, M.; Zawida, Z.; Rusydi, A. H.; Wan, M. K.; Zainudeen, M. H. The Use of Effective Microbes for Worm Controls in Goats–A Preliminary Study. *Malays. J. Vet. Res.*, **2011**, *2*, 57–60.
4. Piccione, G.; Casella, S.; Lutri, L.; Vazzana, I.; Ferrantelli, V.; Caola, G. Reference Values for Some Haematological, Haematochemical, and Electrophoretic Parameters in the Girgentana Goat. *Turk. J. Vet. Anim. Sci.* **2010**, *34*(2), 197–204.
5. Tibbo, M.; Jibril, Y.; Woldemeskel, M.; Dawo, F.; Aragaw, K.; Rege, J. E. O. Factors Affecting Hematological Profiles in Three Ethiopian Indigenous Goat Breeds. *Intern. J. Appl. Res. Vet. Med.* **2004**, *2*(4), 297–309.
6. Olafadehan, O. A. Changes in Haematological and Biochemical Diagnostic Parameters of Red Sokoto Goats Fed Tannin-rich *Pterocarpus erinaceus* Forage Diets. *Vet. Arhiv.* **2011**, *81*(4), 471–483.
7. Zimmermann, I.; Kamukuenjandje, R. T. Overview of Variety of Trials on Agricultural Applications of Effective Microorganism (EM). *Agricola* **2008**, *4*, 17–26.
8. Heita, R. *Supplementing Goats with EM-Bokashi at Kwandu Conservancy.* Report Submitted for National Diploma Course of Agroecology, Polytechnic of Namibia, Windhoek, Namibia, 2006; pp 30.
9. Shiningeni, M. *Control of Intestinal Nematodes in Sheep.* Report Submitted for Bachelor of Technology Research Project, Polytechnic of Namibia, Windhoek, Namibia, 2005.
10. Maisarah, A. Y. Effects of Effective Microbes in Reducing Helminthiasis in Mixed Breed Boer Goats. BSc. diss. Universiti Teknologi MARA, Shah Alam, 2013; pp 38.

11. Esatu, W.; Melesse, A.; Dessie, T. Effect of Effective Microorganisms on Growth Parameters and Serum Cholesterol Levels in Broilers. *Afr. J. Agric. Res.* **2011,** *6*(16), 3841–3846.

12. Greyling, J. P. C. Reproduction Traits in the Boer Goat Doe. *Small Rumin. Res.* **2000,** *36*, 171–177.

13. Najari, S.; Gaddoun, A.; Hamouda, M. B.; Djernali, M.; Khaldi, G. Growth Model Adjustment of Local Goat Population under Pastoral Conditions in Tunisian Arid Zone. *J. Agronomy* **2007,** *6*(1), 61–67.

14. Love., S. C. J.; Hutchinson, G. W. Pathology and Diagnosis of Internal Parasites in Ruminants. In *Gross Pathology of Ruminants, Proceedings 350, Post Graduate Foundation in Veterinary Science*, University of Sydney, Sydney, 2003; Chapter 16, pp 309–338.

15. Aziah. Veterinary officer, Department Veterinary of Selangor, Shah Alam, Selangor. Personal communication, June 4, 2013.

16. Liyana. Effects of Effective Microbes in Reducing Helminthiasis in Goats at Local Farm in Shah Alam. BSc. diss. Universiti Teknologi MARA, Shah Alam, 2013; pp 42.

17. Rinaldi, L.; Veneziano, V.; Morgoglione, M. E.; Pennacchio, S.; Santaniello, M.; Schioppi, M.; Musella, V.; Fedele, V.; Cringoli, G. Is Gastrointestinal Strongyle Faecal Egg Count Influenced by Hour of Sample Collection and Worm Burden in Goats? *Vet. Parasitol.* **2009,** *163*, 81–86.

18. Kumba, F. F.; Katjivena, H.; Lutaaya, E. Seasonal Evolution of Faecal Egg Output by Gastrointestinal Worms in Goats on Communal Farms in Eastern Namibia. *Onderstepoort J. Vet. Res.* **2003,** *70*, 265–271.

19. Houdijk, J. G. M. Influence of Periparturient Nutritional Demand on Resistance to Parasites in Livestock. *Parasite Immunol.* **2008,** *30*, 113–121.

20. Andrew, P.; Krishna, S.; Jacob, A.; Silvina, F.; Andria, J.; Paula, M.; David, K.; America, M.; Alessia, G.; Laura, F.; Bradley, D. W.; John, V. L.; Ralph, M.; Ann, L. B.; Jocelyn, J.; Anita, O. B. *Handbook for the Control of Internal Parasites of Sheep and Goats*. Ontaria Veterinary College, University of Guelph, 2012; pp 15.

21. Basripuzi, H. B.; Sani, R. A.; Ariff, O. M. Anthelmintic Resistance in Selected Goat Farms in Kelantan. *Mal. J. Anim. Sci.* **2012,** *15*, 47–56.

22. Mohyidin, S. Baiduri farm owner, Klang, Selangor. Personal communication, May 24, 2013.

23. Bami, M. H.; Mohri, M.; Seifi, H. A.; Tabatabaee, A. A. A. Effects of Parental Supply of Iron and Copper on Hematology, Weight Gain, and Health in Neonatal Dairy Calves. *Vet. Res. Commun.* **2008,** *32*, 553–561.

24. Qamar, M. F.; Maqbool, A. Biochemical Studies and Serodiagnosis of Haemonchosis in Sheep and Goats. *J. Anim. Plant Sci.* **2012,** *22*(1), 32–38.

25. Torres-Acosta, J. F.; Jacobs, D. E.; Aquilar-Caballero, A.; Sandoval-Castro, C.; May-Martinez, M.; Cob-Galera, L. A. Improving Resilience Against Natural Gastrointestinal Nematode Infections in Browsing Kids During the Dry Season in Tropical Mexico. *Vet. Parasitol.* **2006,** *135*, 163–173.

26. Al-Jebory, E. Z.; Al-Khayat, D. A. Effect of *Haemonchus contortus* Infection on Physiological and Immunological Characters in Local Awassi Sheep and Black Iraqi Goats. *J. Adv. Biomed. Pathobiology Res.* **2012,** *2*, 71–80.

27. Babeker, E. A.; Elmansoury, Y. H. A. Observation Concerning Haematological Profile and Certain Biochemical in Sudanese Desert Goat. *J. Animal Feed Res.* **2013**, *3*(1), 80–86.

28. Elitok, B. Reference Values for Hematological and Biochemical Parameters in Saanen Goats Breeding in Afyonkarahisar Province, *Kocatepe Veterinary J.* **2012**, *5*(1), 7–11.

29. Pambu, R. G. Effects of Goat Phenotype Score on Milk Characteristics and Blood Parameters of Indigenous and Improved Dairy Goats in South Africa. PhD. diss. University of Pretoria, 2011; pp 72.

30. Aluwong, T.; Fatima, H.; Dzenda, T.; Kawu, M.; Ayo, J. Effects of Different Levels of Supplemental Yeast on Body Weight, Thyroid Hormones Metabolism and Lipid Profile of Broiler Chickens. *J. Vet. Med. Sci.* **2012**, 2–15.

31. Krasowska, A.; Kubik, A.; Prescha, A.; Lukaszewicz, M. Assimilation of Omega 6 Fatty Acids and Removing of Cholesterol from Environment by *Saccharomyces boulardii* Strains. *J. Biotechmol.* **2007**, 63–64.

32. Sakha, M.; Shmesdini, M.; Mohamad-Zadeh, F. Serum Biochemistry Values in Raini Goat of Iran. *Internet J. Vet. Med.* **2009**, *6*(1), 1–5.

33. Ikhimioya, I.; Imasuen, J. A. Blood Profile of West African Dwarf Goats Fed *Panicum maximum* Supplemented with *Afzelia Africana* and *Newbouldia laevis*. *Pak. J. Nutr.* **2007**, *6*(1), 79–84.

34. Numushinga, E. *Adding Bokashi to the Diet of Pigs at Drimiopsis*. Report Submitted for Bachelor of Technology Research Project, Polytechnic of Namibia, Windhoek, Namibia, 2005.

35. Aitana, S. *Supplementing Goats and Sheep with EM-bokashi in Omusati Region.* Report Submitted for Bachelor of Technology Research Project, Polytechnic of Namibia, Windhoek, Namibia, 2006

36. Lily Rozita, M. H.; Chandrawathani, P.; Premaalatha, B.; Erwanas, A. I.; Zaini, C. M.; Jmnah, O.; Nurulaini, R.; Nrazura, A. H.; Bohari, J.; Ramlan, M. In *The VRI Small Ruminant Field Programme: Assessment of Parasitic Infections in Local Smallholder Farms*. Proceedings of the 1st ARCAP & 35th MSAP Ann. Conf., 4–6 June 2014, Kuching, Sarawak, Malaysia, 2014; pp 209–210.

CHAPTER 6

RFID APPLICATION DEVELOPMENT FOR A LIVESTOCK MONITORING SYSTEM

M. H. ARIFF[1,*] and I. ISMAIL[2]

[1]*Faculty of Electrical Engineering, Universiti Malaysia Pahang, 26600 Pekan, Pahang, Malaysia*

[2]*Faculty of Electrical Engineering, Universiti Teknologi MARA, 40450 Shah Alam, Selangor, Malaysia*

Corresponding author. E-mail: hisyam@ump.edu.my

CONTENTS

ABSTRACT

Animal rearing is constantly adapting to meet the ever-increasing efficiency and productive demands on the farm field. Moreover, the identification and tracking technologies play an important role in achieving these goals. Radio frequency identification (RFID) has been providing livestock producers with high-quality monitoring systems to deliver continuous control and visibility over automated operations for increased efficiency gains and improved production. This chapter covers a brief idea on livestock monitoring system development with RFID system, reviews the relevant literature, and summarizes current research. We propose a livestock monitoring device which is able to monitor the health status of the livestock using the android smartphone embedded together with the compact UHF RFID reader.

6.1   INTRODUCTION

Ever since the outbreak of the mad cow disease, it has become more important to implement an automatic livestock profiling and tracking, so that the consumers of meat and other animal products can be protected from animal-borne diseases. Radio frequency identification (RFID) plays a vital role by ensuring complete traceability. The meat industry practitioners can now control animal diseases by keeping track of the vaccination data on a regular basis and also checking the health of the cattle. During an outbreak of a disease, the RFID animal identification system can easily identify the flocks affected and these can be isolated to stop further spread of the disease [1].

Animal identification is a big market growing at a rate of over 30% every year [2]. Globally, millions of cattle, pets, and birds are tracked and identified using the electronic identification technologies, primarily RFID technology. Previously, bar codes were used for animal identification, which is now rapidly being replaced by RFID because of the latter's advanced features and advantages. RFID does not require a line-of-sight between the tags and the readers [3]. A reader can read RFID tags from a distance. Furthermore, with anticollision technology, a reader can read multiple tags at a time thus tracking and identifying several animals at the same time. This not just ensures the safety and security of the animals but also reduces the requirement of manual labor and the time taken to identify the animals.

Auto-ID is an automatic identification technology. RFID is grouped under this, along with smartcards, bar codes, and fingerprints or optical recognition systems [4]. RFID is comparable to advanced bar code applications and will continue to work alongside bar codes for the foreseeable future, coexisting until costs reduce and education spreads. The comparison of the two systems is summarized in a Table 6.1.

TABLE 6.1 Comparison Between Bar Code and RFID System.

	Bar Codes	**RFID**
Line of sight	Required	Not required
Technology	Optical	RF waves
Read range	Inches to a few meters	Up to hundreds of meters
Read rates	Slow—single tag can be scanned and read by the reader at a time	Fast—multiple tags can be scanned and read simultaneously
Durability	Exposed—risk of damage, wear and tear during handling	Can be exposed to harsh environment and encased for better protection
Memory capabilities	Static—read only and limited data capacity	Dynamic—read, write, update and high capacity
Security	Low—easily imitated or copied	High—encryption is harder to replicate
Life span	Unlimited—depending on degradation	Up to 10 years
Interference	May be subject to obstruction from dirt or broken from handling	Metal and liquids can interfere with some frequencies
Reusability	No	Yes
Cost	Low	High
Health issues	No	Electromagnetic waves can cause health problems

The main selection in the construction of a data collection system technology is often associated with either the bar code or RFID as listed in Table 6.1. Even though there are similarities, some important differences between the bar code system and RFID technology can be an impact factor that is more suitable to the need of customers. Furthermore, the right solution for their application will depend on whether their situations are better aligned with the capabilities and characteristics associated with each data collection technology. On the other hand, because livestock barn environment is dusty and dirty and also that animal acts are unpredictable, RFID is a wise selection for livestock data collection.

In recent years, the existence of different types of small-sized wearable sensors with powerful functions has facilitated the convenient and automated monitoring of the heart rate and temperature for health monitoring purposes. In addition, nowadays, the introduction of many intelligent smartphones come with smart features such as web browsing, internal database, global positioning system (GPS), wireless local area networks (Wi-Fi), and the Internet. This makes the latest smartphone the best choice for mobile platform development in advanced applications. To improve the project potential, we further study the popular cloud storage platform with a broad range of features. The advantages of cloud storage include linking the information between disease-infected animals and multiple veterinary experts. In addition, it also maintains the livestock report and provides easy access to the livestock information at any time from any location without any obstacles.

6.2 LITERATURE REVIEW

RFID is a modern term used to label a system that wirelessly transmits the identity of an object using radio waves. These transmissions are of unique serial numbers, or codes. This is known as a contactless technology, whereby the tag or item does not need to be manually touched or wired.

The first application used the RFID system in livestock monitoring initially only for identification purposes [5–7]. However, it has gained a growth potential exploration on RFID systems combined with sensors to create other advanced applications in animal rearing sections such as for monitoring the health status of livestock, breeding climate changes, and real-time monitoring on livestock location using GPS [8–10]. In addition, several important general health parameters can be measured for the quick scan of livestock health using sensors such as heart rate, body core temperature, feed intake, head motion, and body weight [11–13].

Nagl et al. [14] designed an implementation of integrated wearable sensor system in livestock health monitoring via Bluetooth communication channels. The purpose of this project is to monitor and record respiration rate, heart rate, and tracking livestock using GPS. All the animal health information can be read through a personal digital assistant (PDA) and wirelessly transmitted from wearable sensors. However, this system is not practical yet, since currently we are engaged with smartphone technology that provides powerful tools compared with PDA.

Kevin et al. [15] designed a livestock health monitoring system using GPS technology and Zigbee module. The Zigbee technology is often used in multiple networks to transmit data for long range and GPS implementation in this project focused on cattle movement for tracing analysis. Although this system is well developed, it is still lacking in terms of database storage medium and still depends on PC-based platform.

While many researchers have published their papers on farm system research and development, little has been done to focus on the latest trends of technology using Android operating system (OS) for livestock health monitoring purposes.

Livestock diseases compromised animal welfare, reduce productivity, and can infect humans. Animal diseases may be tolerated, reduced through animal rearing, or reduced through antibiotics and vaccines. In developing countries, animal diseases are tolerated in animal rearing, resulting in considerable reduced productivity, especially given the low health status of many developing country herds. Disease management for gains in productivity is often the first step taken in implementing an agriculture policy [16]. Disease management can be achieved through changes in animal rearing field monitoring. These measures may aim to control spread using RFID electronic measures, such as controlling animal mixing, controlling entry to farm lots and the use of protective clothing, and quarantining sick animals.

Livestock identification has both visual and management benefits [17]. Identifying an individual livestock with RFID ear tag can make livestock producer easier to determine selected livestock. RFID tags can be classified as passive, active, or semiactive. Active tags have a small battery that pairs with it and periodically sends the signal identification while passive tags do not require any power source to operate. In addition, for semipassive tags, it requires a battery but is not used to improve read range. Table 6.2 shows the comparisons between passive and active tags. Both passive and active tags are being used in livestock monitoring. Passive tags can only provide information on tracking and identification while semipassive or active tags are required for sensing applications.

There are four types of RFID, operating at different radio frequencies as listed in Table 6.3. A business's requirements and its environment will determine which frequency is best to be used. RFID after all is not a simple technology, whereby each system will need modifying, editing, and trial development. RFID frequencies and system zones can be influenced by the environment they are on, with radio waves affected through environmental conditions and things such as metals. There are three common primary RFID frequency bands. Low frequency (LF) is used for implant in trees and animal tracking according to ISO 11784 and ISO 11785 standards. The high frequency (HF) is used for short ranges, up to about 1.5 m reads, with metal and water not affecting the signals. Ultra-high frequency (UHF) offers a better read range with faster reading speeds when compared with LF and HF. However, they use more power and are less likely to pass through the material. The use of super high frequency (SHF) is to avoid interference from metal and water, and thus is practical to be used for climate monitoring and pallet.

TABLE 6.2 Differences Between Active and Passive Tag.

Issues	Active RFID Tags	Passive RFID Tags
Power source	Internal	Energy transfer from the reader
Power availability	Continuous	When reader field is in the range
Operational life	Limited lifetime	Unlimited
Cost	Expensive	Less expansive
Capabilities	Read/write	Read only
Signal strength from tag to reader	High	Low
Signal strength from reader to tag	Low	High
Memory	Battery back SRAM (128 kB or more)	EEPROM (up to 16 kB)
Size	Larger	Smaller
Coverage area	Long distance (>30 m)	Short distance (<10 m)

TABLE 6.3 Features of RFID Tags Based on Frequency.

Frequency Band	LF	HF	UHF	SHF
Common frequency (Hz)	125–135 k	13.56 M	868–928 M	2.45 or 5.8 G
Coupling	Inductive	Inductive	Backscatter	Backscatter
Max read range	<0.5 m	1.5 m	15 m	30 m
Data rate	Low	High	Medium	Medium
Potential uses	Animal identification, item tracking	Animal identification, temperature measuring	Crop monitoring, cold chain monitoring	Pallet, climate monitoring
ISO standards	ISO 14223/1, ISO 18000-2	ISO 14443, ISO 15693, ISO 18000-3	ISO 18000-6	ISO 18000-4
Tag Size	Larger Small			
Energy Efficiency	Higher Lower			

6.3 PROPOSED LIVESTOCK HEALTH MONITORING SYSTEM DESIGN

This research aims to develop a livestock health monitoring device which is able to monitor the health status of the livestock using an Android smartphone. The developed system will be able to communicate with SQL database together with the cloud storage system where livestock producers can have records of information on the livestock heath status or share with other interested parties involved in the livestock field. The development of the architecture of the livestock monitoring system is shown in Figure 6.1.

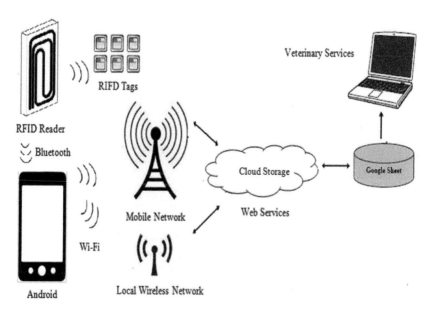

FIGURE 6.1 The architecture of livestock information system.

Furthermore, the livestock health monitoring system will focus on the design and implementation of the mobile UHF RFID tag reader that can be connected via the mini-USB port which is equipped together with an Android smartphone. The details of the prototype are shown in Figures 6.2 and 6.3.

Figure 6.2 shows the concept of UHF reader integrated with mobile smartphones. Moreover, this reader will be designed in a compact size scale for easy use to be put together with a smartphone.

Android Smartphone　　　　　　UHF RFID Reader　　　　　　UHF RFID Tag

FIGURE 6.2　The conceptual design for the livestock information system on mobile.

Figure 6.3 shows the overall picture of the product when it is completed. The UHF RFID reader antenna can be connected to a smartphone via a mini-USB port connectivity.

FIGURE 6.3　Implementation of external UHF RFID reader embeds in smartphone.

In developing an appropriate architecture for data acquisition model, some modules have been developed such as user interface (UI) module, interface module, feedback module, processing module, and database module. All of these modules work together and can be connected to the web services. Other than that, ever since the use of the Android OS in application development, Android components such as activity, intents, services, broadcast receivers, and content providers have been used widely in creating livestock information system. Figure 6.4 shows the data acquisition model of the interconnection between the modules with external systems such as Google sheet facilities and web services.

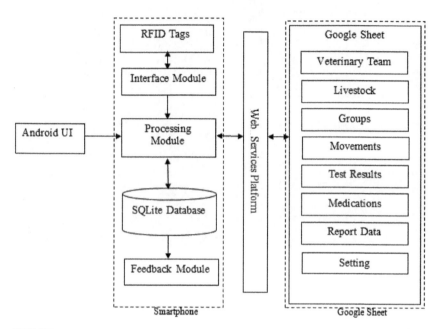

FIGURE 6.4   Data acquisition model.

6.4   PROJECT PLAN

6.4.1   ANDROID APPLICATION DEVELOPMENT

Android phone is a device that operates using the open source Android OS. The manufacturers that produce these devices are Samsung, HTC, Sony, and Motorola. Nowadays, more and more smartphones introduced into the global market are equipped with powerful tools that can do multitasks such as listening to the music while surfing the Internet. The Android software development kit (SDK) is required to build and deploy Android applications. The SDK contains the tools used to test and debug applications. It also contains tools for creating flexible layouts. Figure 6.5 shows the UI menu for livestock health monitoring system. Table 6.4 shows a summary of the main menu and its function.

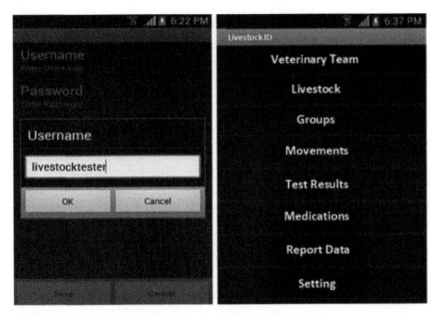

FIGURE 6.5   Graphical user interface for livestock information system.

TABLE 6.4   Summary of Main Menu and Its Function.

Menu	Function
1) Veterinary team	Displays the health care team information such as name, phone number, e-mail address, and address of animal health care center
2) Livestock	Contains record of animals such as sex, date of birth, farm location, status of the livestock, breed, name of breeder, date of infection, and date of disinfections
3) Groups	Contains a summary of information on group animals on the farms such as cattle and sheep
4) Movements	Contains the information of the livestock movement such as source location and transfer new location
5) Test results	Contains the progress report (body temperature, weight, and heart rate) of animal; health graphs on daily, weekly, or monthly
6) Medications	Contains name and quantity of the medicines given to the livestock. The name of caretakers are also included in this menu
7) Report data	Contains specific information of the livestock that can be sent to stockman, breeder, buyer, and the Department of Veterinary Science (DVS)
8) Setting	Setting the connection of RFID reader

6.4.2 UHF RFID READER ANTENNA DEVELOPMENT

The RFID reader antenna construction for this project will be based on the UHF frequency used in Malaysia which ranges from 919 MHz to 923 MHz. In addition, this antenna is developed using a circular polarized (CP) compact patch antenna which is famous for its simple construction, small size, and light weight. The physical parameter of this antenna structure is analyzed using the CST Microwave Studio [18]. Furthermore, all important characteristics for the antenna such as gain, radiation pattern, polarization, and impedance have been taken into account so that the specification of UHF RFID reader installed in smartphones will be achieved. Figure 6.6 shows the ongoing design for UHF antenna reader using CST Microwave Design Studio.

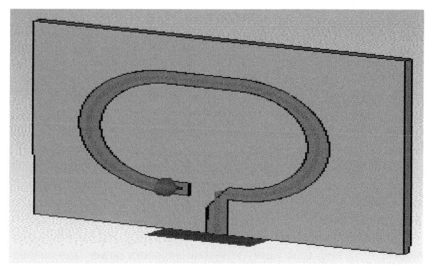

FIGURE 6.6   Ultrahigh frequency design antenna for RFID reader.

6.4.3 DATABASE DEVELOPMENT

In order to store, retrieve, or delete the livestock data, we will design a simple database that link with SQLite database in Android. This data can further be connected to the cloud storage so that any interested parties can access the same information on livestock health. Table 6.4 shows the types of data used for the database development.

6.4.4   FIELD TEST

At this stage, after finalizing the device system integration, it will be tested on the livestock farming owned by the Malaysian Agricultural Research and Development Institute (MARDI) located at Kemaman, Terengganu. The main target of the site analysis is to ensure that the livestock health monitoring system will work in good condition. The following tests will be conducted during the field test, i.e., the receiving test signal, length detection test, and antenna positions to tag detection pattern test for RFID system. Furthermore, for the Android-based applications performance testing, case studies will be performed to validate the whole interaction between the mobile system, wearable sensors, databases, cloud storage, and RFID system.

6.5   CONCLUSION

In this chapter, RFID application development for livestock monitoring system has been presented to determine the health status of the livestock. All the project plan and literature review on this livestock health care device is described. The device is powered by Android operating system provide some information of the heart rate, temperature of livestock, and medication taken by the livestock.

This device integrates several features and functions that make it prominent from others. In most products, determination of livestock information system is based on manual flockbook and some of the devices require complicated setting before doing the measurement.

Finally, this project will be reliable, user friendly, and can be used in the field for a long time since the system generates power from mobile battery system.

ACKNOWLEDGMENT

This research is supported by the Malaysian Ministry of Science, Technology and Innovation Grant 100-RMI/SF 16/6/2 (5/2013). The authors would also like to thank the SIRIM Bhd for providing the facilities to conduct this research and for the financial support throughout the process.

KEYWORDS

- radio frequency identification
- bar code
- livestock monitoring
- ultrahigh frequency
- GPS

REFERENCES

1. Radenkovic, M.; Wietrzky, B. Wireless Mobile Ad-Hoc Sensor Networks for Very Large Scale Cattle Monitoring. In Proceedings of Sixth International Workshop on Applications and Services in Wireless Networks (ASWN'06), Berlin, Germany, 2006, pp 47–58.
2. Frost and Sullivan. Market Inside 2012. Huge Opportunities Lie ahead for UHF RFID in the Growing Animal Identification Market. http://www.frost.com/sublib/display-market-insight-top.do?id=269775891.
3. Patrick, J.; Sweeney, I. *RFID for Dummies*; Wiley Publishing, Inc.: New Jersey, 2005.
4. Ahson, S. A.; Ilyas, M. *RFID Handbook: Applications, Technology, Security, and Privacy*; CRC Press: Boca Raton, 2010.
5. Trevarthen, A. The National Livestock Identification System: The Importance of Traceability in E-business. *J. Theor. Appl. Electron. Commer. Res.* **2007,** *2* (1), 49–62.
6. Ilie-Zudor, E., et al. A Survey of Applications and Requirements of Unique Identification Systems and RFID Techniques. *Comput. Ind.* **2011,** *62*, 227–252.
7. Jones, E. C.; Chung, C. A. *RFID in Logistics: A Practical Introduction*; CRC Press: Boca Raton, 2010.
8. Duan, Y. E. Research on IOT Technology and IOT's Application in Urban Agriculture. *Adv. Mater. Res.* **2012,** *457*, 785–791.
9. Ahuja, L. R.; Ma, L.; Howell, T. A. *Agricultural System Models in Field Research and Technology Transfer*; CRC Press: Boca Raton, 2010.
10. Zerger, A.; Viscarra Rossel, R. A.; Swain, D.; Wark, T.; Handcock, R. N.; Doerr, V.; Bishop-Hurley, G.; Doerr, E.; Gibbons, P.; Lobsey, C. Environmental Sensor Networks for Vegetation, Animal and Soil Sciences. *Int. J. Appl. Earth Obs. Geoinf.* **2010,** *12*, 303–316.
11. Rege, J.; Marshall, K.; Notenbaert, A.; Ojango, J.; Okeyo, A. Pro-poor Animal Improvement and Breeding—What Can Science Do? *Livest. Sci.* **2011,** *136*, 15–28.
12. Kosmider, R. D.; Kelly, L.; Simons, R.; Brouwer, A.; David, G. Detecting New and Emerging Diseases on Livestock Farms Using an Early Detection System. *Epidemiol. Infect.* **2011,** *139*, 1476.

13. Van Metre, D. C.; Barkey, D. Q.; Salman, M.; Morley, P. S. Development of a Syndromic Surveillance System for Detection of Disease Among Livestock Entering an Auction Market. *J. Am. Vet. Med. Assoc.* **2009,** *234*, 658–664.

14. Nagl, L.; Schmitz, R.; Warren, S.; Hildreth, T. S.; Erickson, H.; Andresen, D. Wearable Sensor System for Wireless State-of-Health Determination in Cattle. In *Engineering in Medicine and Biology Society, 2003.* Proceedings of the 25th Annual International Conference of the IEEE, vol. 4, Sept 17–21, 2003, pp 3012, 3015.

15. Smith, K.; Martinez, A.; Craddolph, R.; Erickson, H.; Andresen, D.; Warren, S. An Integrated Cattle Health Monitoring System. In *Engineering in Medicine and Biology Society, EMBS '06. 28th Annual International Conference of the IEEE*, New York, Aug 30, 2006 to Sept 3, 2006, pp 4659, 4662.

16. Thornton, P. K. Livestock Production: Recent Trends, Future Prospects. *Philos. Trans. Royal Soc. B: Biological Sci.* **2010,** *365*, 2853–2867.

17. Barge, P.; Gay, P.; Merlino, P.; Tortia, C. Radio Frequency Identification Technologies for Livestock Management. *Can. J. Anim. Sci.* **2013,** *93*, 23–33.

PART II
Molecular Aspect of Natural Biodiversity

CHAPTER 7

PV92, ACE, AND TPA25 *ALU* INSERTION POLYMORPHISM IN THE KELANTAN MALAYSIA SUB-ETHNIC GROUP

AZZURA ABDULLAH, WAN NURHAYATI WAN HANAFI*, and FARIDA ZURAINA MOHD YUSOF

Faculty of Applied Sciences, Universiti Teknologi MARA, 40450 Shah Alam, Malaysia

Corresponding author. E-mail: wannurhayati@salam.uitm.edu.my

CONTENTS

ABSTRACT

Alu markers have been used extensively for population structure and evolution, both at global and regional level because it permits tracing of population ancestral heritages. The main objective of this study was to determine the *Alu* insertion variability of the Kelantan Malays sub-ethnic group using PV92, ACE, and TPA25 *Alu* loci. Buccal swab samples were obtained from 41 healthy and unrelated subjects of Kelantan Malays. All samples were deposited on FTA paper and directly amplified for PV92, ACE and TPA25. The amplified products were detected and separated via 2% gel electrophoresis. The frequencies of insertion for PV92, ACE, and TPA25 are 0.3415, 0.5121, and 0.4878, respectively. The F-statistics value for all three loci is 0.0229. Analysis of F-statistics had suggested that genetic differentiation existed within Kelantan Malays sub-ethnic group. The incorporation of the information obtained in this study will contribute to basic information regarding PV92, ACE, and TPA25 *Alu* insertion polymorphism variation in Kelantan Malays sub-ethnic group.

7.1 INTRODUCTION

The human genome is one of the most amazing molecular structures in terms of complexity which is ever seen in nature [1]. However, this complex structure can be divided into coding and noncoding regions which were discovered by scientists and which have essential role in human development, physiology, medicine, and phylogeny [1]. The coding regions are responsible for a variety of functions that take place on the DNA and RNA sequences, such as, gene regulation, RNA transcription, RNA splicing, and DNA methylation [2]. The noncoding parts of human genome are composed of various types of repetitive DNA such *Alu* element, which is a member of the short interspersed elements (SINEs) [3]. *Alu* element represents one of the most successful repetitive elements in human genome with over 1 million copies and contributing almost 11% of the genome [3–5]. The majority of these *Alu* elements were inserted into the primate genome 35–60 million years ago, but certain subfamilies of *Alu* elements are relatively very new and suspected to be still evolving [4]. This element mobilized within genome by a gene jumping mechanism known as retroposition, a RNA-mediated transcription process [6]. *Alu* typical sequences are approximately 300 bp nucleotide long. They possess a dimeric structure which is composed of two identical but distinct monomers: the right and left arms are around 200 and 100 nucleotide long, respectively, which are joined together by an A-rich linker and terminated by a short poly (A) tail [7].

Over decades of extensive studies, *Alu* insertions have been found as attractive tools in different fields of scientific research, especially evolutionary, forensic identification, and paternity testing as well as phylogenetic analysis [8]. This is due to several unique and essential properties of this element. As an explanation, once inserted there is no parallel gain or loss of *Alu* element at particular chromosomal location, meaning all chromosomes carrying a polymorphic *Alu* element are identical by descent and inherited from common ancestor [9]. Thus, it will reflect both the maternal and paternal history of a population [9].The ancestral state of these polymorphisms is recognized by the absence of the *Alu* insert [10]. Besides, this element also has homoplasy-free characteristic as the probability of two independent *Alu* insertions occurring in the same genomic region in human population is essentially zero, which causes them to be stable genetic marker [10]. In addition, they are typed by simple, rapid PCR-based assay and do not require radioactive procedures as well as automated DNA sequencers for genotyping [11].

Kelantan, the most northern state of Peninsular Malaysia, borders with Thailand in the north, Perak in the west, Terengganu in the east, and Pahang in the south [12]. This state is well known as "The Beautiful Abode" due to fascinating Palm beaches, scenic fishing villages, traditional costumes, and lifestyle which has been practiced since hundreds of years. Kelantan is also known as the "cradle of Malays culture," due to its highest number of Malays (93%) in Peninsular Malaysia.

Malays in Kelantan are recognized as Kelantan Malays by Malaysian. Kelantan Malays are one out of 14 different Malays sub-ethnic groups (Minang, Bataq, Jambi, Kurinchi, Jawa, Riau, Yunnan, Mendeleng, Banjar, Bugis, Acheh, Champa, Rawa) that are found in Malaysia [13]. According to the federal of constitution of Malaysia, the term Malays refers to person who practices Islam and the Malays culture, who speaks the Malay language, and whose ancestors are Malays [14]. Kelantan Malays differ from other Malays in Malaysia and have preserved their customs, traditions, and cultures well over the years. Some of their most memorable activities are flying giant "Wau" kites and playing "Gasing," a game of spinning tops [15].

7.2   MATERIALS AND METHODS

7.2.1   SUBPOPULATION SAMPLES

In this study, a total of 41 healthy and unrelated individuals of Kelantan Malays were chosen from several districts in Kelantan. Each participant was briefed about the project and was selected based on an interview to ascertain that their pure genetic lineage comprised at least three generation of Kelantan Malays sub-ethnic group. This means both parent and grandparents of each recruited candidate must also originate from Kelantan Malays sub-ethnic group. Those with unknown family history, mixed marriage, and consanguineous marriage were excluded from this study. Prior to the sample collection, the participants were well informed about the aims and objectives of the study and written informed consents were obtained. The ethical approval for this study was obtained from the Research Ethics Committee of Universiti Teknologi MARA (UiTM).

7.2.2   SAMPLE COLLECTION

In this study, the invasive and painless buccal swab collection technique was used to collect buccal cell samples from all 41 unrelated individual Kelantan Malays. The process of sample collection began with earnestly swabbing the inside of a respondent's mouth and rub side by side on the inside cheek by using sterile foam tip for about 1 min. The flat surface of the foam applicator tip was pressed on to the sample circle of fast technology for analysis (FTA) card. The tip was squeezed by using a side to side rocking motion three times to completely saturate the sample area. The change of FTA card from pink to white indicated the presence of cells. The FTA cards were allowed to dry and stored at room temperature which were then ready for further use.

7.2.3   PCR AMPLIFICATION

Three human-specific *Alu* makers (PV92, ACE, TPA25) were typed in this study. They are commonly used in worldwide study of human population genetics [5]. Details of the selected oligonucleotide primers are outlined in Table 7.1. Amplification of *Alu*-inserted regions was carried out in

BIOER Xpcycler. PCR was performed in a total volume of 50 μL reaction consisting 2-mm FTA cards, 2x Phusion Human Specimen PCR Buffer, 10 mM oligonucleotide primer, Phusion Human Specimen DNA polymerase, and sterile H_2O. The PCR protocol comprised of an initial denaturation of 5 min at 98°C followed by 35 amplification cycles of denaturation at 98°C for 1 s, optimized annealing temperature for 30 s, and followed by extension at 72°C for 15 s. The final extension temperature was set at 72°C for 1 min, followed by the holding temperature at 4°C, and the final product was stored at 4°C. Subsequently, 10-μL amplified products were resolved on 2% (w/v) agarose gel containing 1.3 μL Gold View™ Nucleic Acid stain. PCR products were directly visualized under UV transilluminator and molecular weight was determined using 100-bp DNA ladder. All DNA samples were kept for further use.

TABLE 7.1 Oligonucleotide Primers and PCR Conditions for *Alu* Loci Studied.

Alu Locus	Fragment Sizes, bp	Sequence	Annealing Temperature, Ta (°C)
PV92	416/101	F–5'_AACTGGGAAAATTTGAAGAGAAAGT_3'	59.2
		R–5'_TGAGTTCTCAACTCCTGTGTGTTAG_3'	
ACE	490/190	F–5'_CTGGAGACCACTCCCATCCTTTCT_3'	53.9
		R–5'_GATGTGGCCATCACATTCGTCAGAT_3'	
TPA25	457/134	F–5'_GTAAGAGTTCCGTAACAGGACAGCT_3'	55
		R–5'_CCCCACCCTAGGAGAACTTCTCTTT_3'	

7.2.4 DATA ANALYSIS

Allele frequencies were calculated using the gene counting method according to the formula given below:

$$\text{Frequency of an allele } (p) = \frac{2 \times \text{number of homozygotes } (+, +) + \text{heterozygotes}}{2 \times \text{total number of individuals } (N)},$$

$$\text{Frequency of an allele } (q) = \frac{2 \times \text{number of homozygotes } (-, -) + \text{heterozygotes}}{2 \times \text{total number of individuals } (N)}.$$

Using these allele frequencies, heterozygosity at individual locus was estimated according to the formula $H = 2pq$, where p and q are the frequency of an allele (+) and allele (−), respectively. Finally, F-statistics was calculated to analyzed variation in the gene frequencies.

7.3 RESULTS AND DISCUSSION

Figure 7.1 shows the photograph of 2% (w/v) agarose gel containing the PCR product of ACE *Alu* insertion polymorphism in Kelantan Malays. The length of PCR product including the priming sites and flanking DNA are approximately 190 bp to 490 bp. These PCR products are successfully amplified at annealing temperature of 53.9°C. From extreme left, lane 1 belongs to 100 bp DNA marker that functions to determine the size of the PCR product by comparing the migration of the band distance with the size of the DNA marker. Lane 3, 5, 6, 16, and 17 possess homozygous individuals (+,+) of ACE *Alu* insertion polymorphism as the presence of single 490-bp PCR product. The heterozygous individuals (+,−) are presented in lane 9 with the presence of double band of 190 bp and 490 bp PCR product. The homozygous individuals for the absence (−,−) of ACE *Alu* insertion polymorphism are indicated in lane 4, 7, 8, 10, 12 until 15, 18, and 19 as the presence of single band at 190 bp.

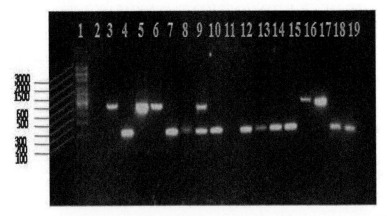

FIGURE 7.1   Photograph of a 2% agarose gel containing the PCR products of *Alu* ACE. Lane from extreme left: 1: 100 bp DNA marker; 2 and 11: No sample loaded; 3, 5, 6, 16 and 17: Homozygous individuals (+,+); 9: Heterozygous individual (+,−); 4, 7, 8, and 10: Homozygous individuals (−,−); 12 until 15, 18 and 19: Homozygous individuals (−,−).

Table 7.2 represents the successful amplification of all three *Alu* loci (PV92, ACE, TPA25) in 41 Kelantan Malays sub-ethnic group. For PV92 it demonstrates that there are 28 individuals who are heterozygous (+,−) and other 13 are homozygous for the absence (−,−) of *Alu* insertion polymorphism. Besides for ACE *Alu* locus, there are nine individuals who are homozygous for the insertion (+,+), 24 individuals are heterozygous (+,−), and the balance of eight are homozygous for the absence (−,−) of ACE. Furthermore, there are also 12 and 16 individuals that, respectively, show homozygous insertion (+,+) and heterozygous insertion (+,−) of TPA25 *Alu* insertion polymorphism. In addition, 13 out of 41 individuals show that they are homozygous for the absence of TPA25 (−,−) *Alu* insertion polymorphism.

TABLE 7.2 Amplification of PV92, ACE, and TPA25 *Alu* Loci.

Individual	Genotype			Individual	Genotype		
	PV92	**ACE**	**TPA25**		**PV92**	**ACE**	**TPA25**
1	(−,−)	(−,−)	(+,−)	22	(+,−)	(−,−)	(+,−)
2	(−,−)	(+,+)	(+,+)	23	(+,−)	(+,+)	(−,−)
3	(+,−)	(+,+)	(+,−)	24	(+,−)	(+,−)	(+,+)
4	(−,−)	(+,−)	(+,−)	25	(+,−)	(+,−)	(−,−)
5	(+,−)	(+,−)	(+,−)	26	(−,−)	(+,−)	(+,−)
6	(+,−)	(−,−)	(+,+)	27	(−,−)	(+,−)	(−,−)
7	(+,−)	(+,−)	(−,−)	28	(+,−)	(+,−)	(−,−)
8	(+,−)	(+,+)	(+,+)	29	(+,−)	(+,−)	(−,−)
9	(+,−)	(+,+)	(+,+)	30	(+,−)	(+,−)	(−,−)
10	(+,−)	(+,+)	(−,−)	31	(−,−)	(+,−)	(+,+)
11	(+,−)	(+,−)	(+,+)	32	(−,−)	(+,−)	(+,+)
12	(+,−)	(+,+)	(+,−)	33	(−,−)	(+,−)	(−,−)
13	(+,−)	(+,−)	(+,−)	34	(−,−)	(+,+)	(+,−)
14	(+,−)	(+,+)	(+,−)	35	(−,−)	(+,−)	(+,−)
15	(+,−)	(+,−)	(+,−)	36	(−,−)	(+,−)	(−,−)
16	(+,−)	(+,−)	(+,+)	37	(+,−)	(+,−)	(−,−)
17	(+,−)	(+,−)	(+,−)	38	(−,−)	(−,−)	(−,−)
18	(+,−)	(+,−)	(+,−)	39	(−,−)	(−,−)	(+,+)
19	(+,−)	(+,−)	(+,−)	40	(−,−)	(−,−)	(+,+)
20	(+,−)	(−,−)	(+,+)	41	(+,−)	(−,−)	(+,+)
21	(+,−)	(+,−)	(−,−)				

Table 7.3 shows allele frequencies and heterozygosity of all three *Alu* loci in Kelantan Malays sub-ethnic group. The frequencies of inserted allele (+) were 0.3415, 0.5121, 0.4878 for PV92, ACE, and TPA25, respectively. None of these markers were completely fixed in Kelantan Malays sub-ethnic group. The higher deletion frequencies belongs to PV92 (0.6585), followed by TPA25 (0.5122) and ACE (0.4879).

TABLE 7.3 Allele Insertion Frequencies at Three *Alu* Loci.

	Kelantan Malays, $N = 41$		
Locus	**PV92**	**ACE**	**TPA25**
Allele frequencies			
Insertion	0.3415	0.5121	0.4878
Deletion	0.6585	0.4879	0.5122
Heterozygosities			
Observed	0.6829	0.5854	0.3902
Expected	0.4498	0.4997	0.4997

Table 7.4 shows the F-statistics of all three *Alu* loci that were examined for Kelantan Malays sub-ethnic group. From the table, we can observe that the value of F_{IS} is negative for all three loci in this sub-ethnic group (−0.1486). The role of F_{IS} in a population study is to determine the reduction in heterozygosity of an individual that results from nonrandom mating within its population. The negative value of F_{IS} for PV92, ACE, and TPA25 loci indicates that there were excess of heterozygotes in these loci. In addition, there was also negative value for F_{IT} (−0.1223) which indicates that there was deficiency of homozygote at each of three polymorphic loci for this Malays sub-ethnic group. As the value of F_{ST} (0.0229) is considered

TABLE 7.4 The F-Statistics for Three *Alu* Loci in Kelantan Malays Sub-ethnic Group.

	Kelantan Malays, ($N = 41$) (PV92, ACE, TPA25)
H_I	0.5549
H_S	0.4831
H_T	0.4944
$F_{IS} = 1 - (H_I/H_S)$	−0.1486
$F_{IT} = 1 - (H_I/H_T)$	−0.1223
$F_{ST} = 1 - (H_S/H_T)$	0.0229

low, it represents that there is genetic differentiation occurring within Kelantan Malays sub-ethnic group.

Malaysia is one of most unique countries that is renowned all around the world. Malaysians are composed of three major ethnic groups, namely, Malays, Chinese, and Indian. These three major ethnic groups can be easily differentiated via their physical characteristic like skin color, eyes shape, and the language that they use [16]. However the Malays which are the majority of Malaysians in this country (63.1%) can be divided into other 14 small sub-ethnic groups (Kelantan, Minang, Bataq, Jambi, Kurinchi, Jawa, Riau, Yunnan, Mendeleng, Banjar, Bugis, Acheh, Champa and Rawa) [13]. All these sub-ethnic groups can be found in several particular locations in Malaysia. The most important is that, all these sub-ethnic groups show some similarity in terms of physical characteristic, but they have their own particular language, food, and culture. As the name, Kelantan Malays can be found mostly in Kelantan. Compared with other Malays sub-ethnic groups, they can be recognized from the way they communicate, the words they use, the food they consume, dressing, and also traditional game [15]. For example, Kelantan Malays men prefer to wear their traditional Baju (loose shirt top), with Seluar (long pants) and a Sarung wrapped around their midriff in their daily activity [17]. Besides, they also communicate via their own language that differs from Bahasa Melayu or other Malays sub-ethnic groups' language [18]. Several simple words are used, such as Tubik = going out, Pitih = money, and Samah = RM 0.50 cents [19]. In addition, this Malays sub-ethnic group has their very unique traditional game known as kite flying Wau Bulan (moon-kite) and Gasing (top-spinning) [15]. There are 14 Malays sub-ethnic groups identified in Malaysia; however, no data on DNA *Alu* insertion polymorphism as genetic markers have been reported on Kelantan Malays sub-ethnic group in the literature.

As the Kelantan state is located in the South East Asia region, we have compared the frequencies insertion of *Alu* PV92 and TPA25 to a nearest population like Cambodian, Chinese, Japanese, and Vietnamese. However, Kelantan Malays sub-ethnic groups did not show any similarities with all these populations. This is because all these populations possess high insertion of PV92 with range (0.8529–1.000) [20]. The frequencies of TPA25 also contrast due to higher insertion in Cambodian (0.5417), Chinese (0.4412), Japanese (0.5000), and lower in Vietnamese (0.2778) [20]. Furthermore, the frequencies of ACE insertion are also high in Java (0.86) and Taiwan (0.50) [21] compared with Kelantan Malays sub-ethnic group.

Nevertheless, Kelantan Malays show some similarity toward PV92 *Alu* insertion frequency to Gujjar population (0.3435) that is located in North West India [8], North Morocco (0.3330), and West Morocco (0.3430) [22]. In addition, the insertion frequency of ACE is also close to Central Bosnia (0.5000), Middle Bosnia (0.5000) [9], and Tata population (0.5150) [23] of South Morocco. Finally, the TPA25 insertion frequencies have been observed to show some similarities with Tata (0.4850) and an Indian population known as Kapu (0.4818) [24].

7.4 CONCLUSION

The incorporation of the information obtained in this study will contribute to provide basic information regarding PV92, ACE, and TPA25 *Alu* insertion polymorphism variation in Kelantan Malays sub-ethnic group. This effort will help to start *Alu*-based information of Malays sub-ethnic groups that other researchers can rely on that may include: creating secondary database, comparative genomics for related problems, or even for forensic purpose. Besides, this study will help to enhance the growth of Bioinformatics research among research community in Malaysia. Further studies will be conducted on more *Alu* loci to fully understand the nature and extent of genetic differentiation among Kelantan Malays sub-ethnic group.

ACKNOWLEDGMENT

This study was supported by Ministry of Sciences Technology and Innovation Grant (02-01-01-SF0639).

KEYWORDS

- **Kelantan Malays**
- ***Alu* markers**
- **PCR**
- **allele frequency**
- **F-statistic**

REFERENCES

1. Moreno, P. A.; Vélez, P. E.; Martínez, E.; Garreta, L. E.; Díaz, N.; Amador, S.; Tischer, I.; Gutiérrez, J. M.; Naik, A. K.; Tobarand, F.; García, F. The Human Genome: A Multifractal Analysis. *BMC Genom.* **2011**, *12*, 1–17.
2. Butler, J. M. *Fundamentals of Forensic DNA Typing*, 1st ed.; Academic Press, 2009; p 350.
3. Deininger, P. *Alu* Element Know the Sines. *BMC Genom.* **2011**, *12*, 1–12, 2011.
4. Ahmed, M.; Liand, W.; Liang, P. Identification of Three New Alu Yb Subfamilies by Source Tracking of Recently Integrated Alu Yb Elements. *Mobile DNA* **2013**, *4*, 1–11.
5. Solovieva, D. S.; Balanovska, E. V.; Kuznetsova, M. A.; Vasinskaya, O. A.; Frolova, S. A.; Pocheshkhova, E. A.; Evseeva, I. V.; Boldyreva, M. N.; Balanovsky, O. P. The Russian Gene Pool: The Gene Geography of Alu Insertions (*ACE, APOA1, B65, PV92, TPA25*). *Mol. Biol.* **2010**, *44*, 393–400.
6. Kee, B. P.; Chua, K. H.; Lee, P. C.; Laian, L. H. Population Data of Six *Alu* Insertions in Indigenous Groups from Sabah, Malaysia. *Ann. Hum. Biol.* **2012**, *39*, 505–510.
7. Dridi, S. *Alu* Elements: From Junk DNA to Genomic Gems. *Scientiica* **2012**, *2012*, 1–11.
8. Saini, J. S.; Kumar, A.; Matharoo, K.; Sokhi, J.; Badaruddoza; Bhanwer, A. J. S. Genomic Diversity and Affinities in Population Groups of North West India: An analysis of Alu Insertion and a Single Nucleotide Polymorphism. *Gene* **2012**, *511*, 293–299.
9. Pojskic, N.; Silajdzic, E.; Kalamujic, B.; Kapur-Pojskic, L.; Lasic, L.; Tulic, U.; Hadziselimovic, R. Polymorphic Alu Insertion in Human Populations of Bosnia and Hezergovina. *Ann. Hum. Biol.* **2013**, *40*, 181–185.
10. Akhatova, F. S.; Boulygina, E. A.; Litvinov, S. S.; Rizvanova, F. F.; Khusnutdinova, E. K. Analysis of Polymorphism in Alu-Incertional Loci in the Population of Tatar in Russia. *World Appl. Sci. J.* **2013**, *24*, 461–463.
11. Yadav, A. S.; Arora, P. Genomic Diversities and Affinities among Eight Endogamous Groups of Haryana (India): A study on Insertion/Deletion Polymorphisms. *Ann. Hum. Biol.* **2011**, *38*, 114–118.
12. Mahirah, K.; Khalid, A. R. Assessing Consumer's Willingness To Pay for Improved Domestic Water Services in Kelantan, Malaysia. *Int. J. Sci.* **2013**, *8*, 45–53.
13. Peng, H. B.; Nur Shafawati, A. R.; Gin, O. K.; Zilfalil, A. Y-chromosomal STR Variation in Malays of Kelantan and Minang Pertanika. *J. Trop. Agric. Sci.* **2008**, *31*, 135–140.
14. Rajadurai, J. Speaking English and the Malays Community. *Indones. Malays World* **2010**, *38* (11), 289–301.
15. Aziz, S. A. The Kelantan Traditional Arts as Indicators for Sustainability: An Introduction to Its Genius Loci. *J. Soc. Sci.* **2013**, *2*, 41–54.
16. Nakamura, R. Malaysia, a Racial Nation: Study of the concept of Race in Malaysia. *Int. Proc. Econ. Dev. Res.* **2012**, *42*, 134.
17. Hamzah, A.; Ismail, H. N. *A Design of Nature-Culture Based Tourism Corridor; A Pilot Project at Kelantan Darul Naim;* eprints. Universiti Sains Malaysia, 2008, pp 1–107.

18. Hasniza, C.N.C. *Dialek Kelantan vs Bahasa Melayu: Tinjauan dari perspektif komunikasi silang budaya*; Pustaka. UPSI. Universiti Pendidikan Sultan Idris, Malaysia, 2010, pp 1–13.

19. Mohamad Nazri, A.; Hanapi, D.; Seong, T. K. Penghayatan Tradisi Lisan Melayu Dalam Kalangan Masyarakat Cina Kelantan. *Jurnal Melayu* **2011,** *8,* 111–132.

20. Watkins, W. S.; Ricker, C. E.; Bamshad, M. J.; Carroll, M. L.; Nguyen, S. V.; Batzer, M. A.; Harpending, H. C.; Rogers, A. R.; Jorde, L. B. Patterns of Ancestral Human Diversity: An Analysis of *Alu*-Insertion and Restriction-Site Polymorphisms. *Am. J. Hum. Genet.* **2001,** *68,* 738–752.

21. Stoneking, M.; Fontius, J. J.; Clifford, S. L., et al. *Alu* Insertion Polymorphism and Human Evolution: Evidence for a Larger Population Size in Africa. *Genome Res.* **1997,** *7,* 1061–1071.

22. Cherni, L.; Frigi, S.; Ennafaa, H.; Mtiraoui, N.; Mahjoub, T.; Benammar-Elgaaied, A. Human *Alu* Insertion Polymorphisms in North African Populations. *Hum. Biol.* **2011,** *83,* 611–626.

23. Chadli, S.; Wajih, M.; Izaabel, H. Analysis of *Alu* Insertion Polymorphism in South Morocco (Souss): Use of Markers in Forensic Science. *Open Forensic Sci. J.* **2009,** *2,* 1–5.

24. Romualdi, C.; Balding, D.; Nasidze, I. S.; Risch, G.; Robichaux, M.; Sherry, S. T.; Stoneking, M.; Batzer, M. A.; Barbujani, G. Pattern of Human Diversity, Within and among Continent, Inferred from Biallelic DNA Polymorphims. *Genom. Res.* **2002,** *12,* 602–612.

CHAPTER 8

SCREENING AND ISOLATION OF LOCALLY THERMOPHILIC FACULTATIVE ANAEROBE BACTERIA FOR BIOSURFACTANT PRODUCTION

NURUL FATIHAH[1,*], T. Z. M. TENGKU ELIDA[1], A. K. KHALILAH[1,] W. O. WAN SITI ATIKAH[2], S. SABIHA HANIM[3], and A. AMIZON[4]

[1]*Department of Biology, Faculty of Applied Sciences, Universiti Teknologi MARA, 40450 Shah Alam, Malaysia*

[2]*Department of Biology, Faculty of Applied Sciences, Universiti Teknologi MARA, 26400 Bandar Jengka, Malaysia*

[3]*Department of Chemistry, Faculty of Applied Sciences, Universiti Teknologi MARA, 40450 Shah Alam, Malaysia*

[4]*Faculty of Chemical Engineering, Universiti Teknologi MARA, 40450 Shah Alam, Malaysia*

Corresponding author. E-mail: nurulfatihahkhairuddin@yahoo.com

CONTENTS

ABSTRACT

Biosurfactants are well known for their biodegrading potential and low hazard to the environment which explains why they have very promising applications environmentally and biotechnologically. Majority of the studies reported were using mesophilic bacterial strains but this may cause problems as these strains were not able to survive after being applied to oil spills having a temperature above 50°C. Thermophilic bacterial species are able to grow on a hydrocarbon-containing medium at a temperature up to 50°C and suitable for use in microbial-enhanced oil recovery (MEOR) and oil-sludge clean-up. A thermophilic *Bacillus* sp. was isolated from hot spring Sungai Klah, Perak, Malaysia, and later identified as *Anoxybacillus* sp. using biochemical tests and 16S rRNA sequence analysis. *Anoxybacillus* sp. is a new genus compared with the well-studied *Geobacillus* sp. or *Bacillus* sp. The strain was screened as positive biosurfactant producer and the stability of biosurfactant was characterized under the influence of different range of temperature, pH, and salinity. The surface activity of the sample relatively remained stable between pH 10 and 12 indicating preference for alkaline conditions. At the temperature range 4–121°C, the surface tension activity of the biosurfactant was maintained and did not show any significant loss of activity but the most stable condition was at temperature 25°C (33.97 mN/m). The surface tension started to decline at 4% to 6% (w/v) sodium chloride but was stable within this range. Based on the stability of the biosurfactant under the influence of different range of temperature, pH, and salinity, biosurfactant showed promising potential in environmental cleaning applications.

8.1   INTRODUCTION

Microbial surfactants or commercially known as biosurfactants are biomolecules that are synthesized by a variety of microorganisms. It is usually synthesized under specific growth conditions either on water miscible or oily substrate [21]. Biosurfactants are amphiphilic molecules that have two domains which are hydrophobic and hydrophilic region. The accumulations of these molecules at the interface will induce the formation of micelles which can lead to the reduction of surface and interfacial tensions. It will enhance the solubility and mobility of the insoluble or hydrophobic compounds [1, 21]. Recently, biosurfactant gained importance in various commercial applications in biological industries, environmental protection, food processing, pharmaceuticals, biomedical, cosmetics, and agricultural industries. Biosurfactants have huge potential to replace synthetic (chemically produced) surfactants. Unlike most synthetic surfactants, many biosurfactants function effectively at extremes of temperature, salinity, wide range of pH, low toxicity, better foaming (useful in mineral processing), and are environment friendly in nature [19].

Majority of the studies reported were using mesophilic bacterial strains but studies using thermophilic bacterial strains are rare [22]. A thermophilic biosurfactant producer, *Bacillus* sp., was able to grow on a hydrocarbon-containing medium at temperature up to 50°C and suitable for use in microbial-enhanced oil recovery (MEOR) and oil-sludge cleanup [2]. *Anoxybacillus* sp. produced biosurfactant but studies about producing biosurfactant using this genus are rare [14, 24]. The genus *Anoxybacillus* belongs to the class *Bacilli*, order *Bacillales*, family *Bacillaceae*, and under the *Firmicutes* phylum in the bacterial domain. The first strict anaerobic *Anoxybacillus* sp., *Anoxybacillus pushchinensis*, was isolated from manure [15]. In addition to *A. contaminans* which was isolated from contaminated gelatine from a manufacturing plant, other newly described species have originated from various geothermal sites around the globe [7]. Examples of these species include *A. flavithermus*, *A. gonensis*, *A. ayderensis*, *A. kestanbolensis*, and *A. amylolyticus* [17].

The aim of the present study was to isolate and characterize the main functional properties of biosurfactant produced by isolated thermophilic bacterial strain. Characterization included the determination of stability effect of biosurfactant against different factors such as temperature, pH, and salinity. The identification was also studied to confirm the potential species of the isolated biosurfactant producer.

8.2   MATERIALS AND METHODS

8.2.1   SAMPLE COLLECTION AND CULTURE ENRICHMENT

The sample was collected from a natural hot spring located in Sungai Klah, Tanjung Malim, Perak, Malaysia. The sampling strategy was reported previously by Hussin et al. [10]. The temperature was 55°C and the pH was 7.6. One milliliter of water and sediment mixture from hot spring were inoculated into 100 mL of nutrient broth in 250 mL Erlenmayer flask and shaken (150 rpm) at temperature 55°C for 24–48 h.

8.2.2 THERMOPHILIC BACTERIAL ISOLATION

One milliliter of enrichment culture was inoculated into 100 mL of minimal salt medium (MSM) in 250 mL Erlenmayer flask supplemented with 1 mL trace element and 2% (v/v) diesel was used as the sole carbon source and shaken (150 rpm) at temperature 55°C for 5 days [11]. Samples were serially diluted and plated by spread plate method on nutrient agar covered with diesel and incubated at temperature 55°C for 24 h. The composition of the MSM (g/L): KH_2PO_4-0.2; K_2HPO_4-0.3; $MgSO_4.7H_2O$-0.5; $CaCl_2$-0.15; $NaCl_2$-0.5; $NaNo_3$-1[8]. The composition of trace element was (mg/l): $ZnSO_4.7H_2O$-50; $MnCl_2.4H_2O$-400; $CoCl.6H_2O$-1; $CuSO_4.5H_2O$-0.4; H_3BO_2-2; $NaMoO_4.2H_2O$-50

8.2.3 BIOSURFACTANT SCREENING

Hemolytic activity was carried out as described by Carillo et al. [6]. Isolated strains were screened on blood agar plates containing 5% (v/v) blood and incubated at temperature 55°C for 24–48 h. Hemolytic activity was indicated in the form of the presence of a clear zone around a colony.

Drop-collapse test was performed on clean glass slides [4]. The glass slides were rinsed with hot water followed by ethanol and distilled water, and dried. The slides were then coated with 1.5 µL of crude oil and left to stand for equilibration for 24 h. A 5-µL aliquot of sample was then applied onto the center of the oil drops. The results were monitored visually after 1 h.

Oil spreading technique was carried out in the disposable petri dish as described by Youssef et al. [26] and Jaysree et al. [12]. Fifteen milliliter of distilled water was added to petri dishes followed by addition of 100 µL of crude oil onto the water surface. Ten microlitre of cell culture supernatant was later dropped onto crude oil surface. The diameter of the clear zone on the oil surface was measured.

The emulsifying activity was determined using the method described by Willumsen and Karlson [25]. Two microlitre of supernatant obtained from centrifugation at 13,000 rpm was added to different hydrocarbons of the same volume such as kerosene and petrol in different test tubes and mixed for 2 min by using a vortex. The emulsification index, E24 (%), is the ratio of the height of emulsified layer (mm) divided by the total height of the liquid mixture (mm) and multiplied by 100 (E24 = a/b × 100). An emulsion was defined as stable if the E24 was 50% or better [5].

Surface tension was measured with pendant drop shape technique using pendant drop analyzer at room temperature.

All the experiments were carried out in triplicates.

8.2.4 STABILITY STUDY

The stability of biosurfactant obtained from the selected isolate was studied under different range of temperature (55°C→25°C, 25°C→4°C, 25°C→70°C, 25°C→100°C, 25°C→121°C→25°C, 25°C→121°C→4°C), pH (2, 4, 6, 8, 10, 12), and salinity [2%, 4%, 6%, 8%, and 10% (w/v)] and the surface tension was also determined.

8.2.5 IDENTIFICATION OF THE THERMOPHILIC BIOSURFACTANT PRODUCER

Morphological and biochemical tests were performed according to Bergey's Manual of Systematic Bacteriology [18]. Partial sequencing of 16S rRNA was carried out in First Base Laboratories, Malaysia. The 16S rRNA gene nucleotide sequences were compared with known sequences in GenBank using BLAST program.

8.2.6 EXTRACTION OF BIOSURFACTANT

Biosurfactant was extracted from the culture supernatant after cell removal by centrifugation at 8000 g for 15 min in a centrifuge (Heraeus Biofuge) at 4°C. The supernatant was adjusted to pH 2 using 1 M sulphuric acid and incubated overnight at 4°C. Grey precipitate was collected by centrifugation at 8000 g for 15 min then extracted with 10 mL chloroform–methanol mixture (2:1, v/v). The mixture was shaken for 15 min at 30°C and 250 rpm. The content was then centrifuged (Heraeus Biofuge) at 8000 g for 15 min at 4°C and the supernatant was evaporated by air drying. The remaining residue was dispensed in sodium phosphate buffer (pH 7.0) and stored at 4°C.

8.2.7 FOURIER TRANSFORM INFRARED SPECTOSCOPY

Fourier transform infrared spectroscopy (FT-IR) was used to determine the composition of an unknown mixture. One milligram of purified sample was ground with 100 mg of KBr and compressed to obtain translucent pellets. The infrared absorption spectra were measured at spectral resolution and wave number accuracy of 4 and 0.01 cm^{-1}, respectively. KBr pellet was used as background reference. The spectra were compared with the database to confirm the chemical nature of the active biosurfactant fraction.

8.3   RESULTS AND DISCUSSION

8.3.1 ISOLATION AND SCREENING OF POTENTIAL BIOSURFACTANT PRODUCER

A total of 10 isolates (A, B, C, D, E, F, G, H, I, and J) were isolated from hot spring under thermophilic condition (Table 8.1). The isolates were later screened for the production of biosurfactants when grown in diesel oil at 55°C. All of them were spore forming with spore located at terminal of the cell and Gram positive except isolates B and F. They were rod in shape and motile. All colonies were round in shape with entire margin and rise elevation.

All the isolates were screened using qualitative and quantitative screening methods [23]. The first method was used to screen biosurfactant producer based on the lysis of blood cells on blood agar. Carrillio et al. [6] found an association between hemolytic activity and the production of surfactant. They recommended using this method as a preliminary screening for biosurfactant activity [2, 12, 13, 22]. Only two isolates (D and I) gave positive results for blood hemolysis in this study. Youssef et al. [26] reported 13.5% of the hemolytic strains reduced the surface tension to values below 40 mN/m. Many authors suggested that this method should be supported by other techniques based on the surface activity measurement [12, 13] since it can produce a lot of false negative and false positive results due to several reasons: effect of lytic enzymes that may be produced by the bacteria, hydrophobic substrates should not be included as sole carbon source in this assay, and diffusion restriction of the surfactant might inhibit the formation of clearing zones [23]. In addition, Schulz et al. [20] showed that some biosurfactants do not show any hemolytic activity at all. This method excluded many efficient biosurfactant producers. Some authors note that its use is limited to screen for microorganisms that produce water-soluble, diffusible, low molecular weight biosurfactants during growth [3].

TABLE 8.1   Screening of Thermophilic Biosurfactant Producer by Selected Methods: Blood Lysis, Drop Collapse, Oil Spreading (Os), Emulsification (E24), and Surface Tension (ST).

Isolate	Blood Lysis	Drop Collapse	OS (mm)	E24 Kerosene (%)	E24 Crude Oil (%)	ST (mN/m)
A	−	+	16.50	13.50	20.00	34.58
B	−	+	3.50	16.50	16.50	37.50
C	−	−	3.00	10.00	10.00	45.68
D	+	−	2.00	10.00	13.00	47.48
E	−	−	3.00	10.00	13.00	42.50
F	−	−	3.00	10.00	13.00	42.75
G	−	−	3.00	10.00	13.50	43.88
H	−	−	3.00	10.00	13.50	42.13
I	+	−	3.00	10.00	13.00	47.43
J	−	−	1.50	10.00	13.50	46.13
+ [a]		+	5.50	67.00	60.00	26.05
− [b]		−	0.00	0.00	0.00	54.12

[a]Positive control.
[b]Negative control.

Youssef et al. [26] suggested a simple protocol for screening of biosurfactant producer if it involves a large number of microorganisms. Initially, the isolates should be analyzed using the drop-collapse method as a presumptive test. Isolates with negative results were later screened using oil spreading technique to detect those that produce low level of biosurfactants. Finally, the surface tension measurements were carried out for confirmation. Both the oil spreading and drop-collapse techniques can be used as qualitative and quantitative assays. Moreover, they have several advantages due to these techniques since it requires a small volume of sample, rapid and easy to carry out, and not requiring specialized equipment. Two isolates (A, B) showed positive results for biosurfactant production using both oil spreading technique and drop-collapse methods, and only eight isolates showed positive for biosurfactant production using oil spreading technique. All isolates showed reduced surface tension values ranging from 34.58 to 47.48 mN/m. The results obtained indicated that the oil spreading technique is more reliable for detecting biosurfactant compared to drop-collapse method since it is more sensitive to low concentration of biosurfactant. It is also much easier to carry out and is less time consuming.

In this study, the protocol proposed by Youssef et al. [25] was applied to screen thermophilic biosurfactant-producing bacteria with additional method employed, that is, emulsification activity. The emulsification abilities obtained using E24 were presented in Table 8.2. The values of E24 in the presence of crude oil ranged from 10% to 20%, whereas the values of E24 in the presence of kerosene ranged from 10% to 16.5%. The values of E24 of isolates A and B were quite low although they had high surface tension reduction ability compared to other isolates. Plaza et al. [16] reported that the bacterial strains that had high surface tension reduction were not generally able to form the emulsions. Willumensen and Karlson [24] also reported similar findings that further confirmed no correlation existed between reduction of surface tension and formation of emulsion.

TABLE 8.2   Surface Tension (ST) Activity at Different Temperatures.

Temperature Changes (°C)	Surface Tension (mN/m)
55–25	33.97
25–4	34.69
25–70	34.81
25–100	35.03
25–121–25	34.52
25–121–4	34.82

8.3.2   IDENTIFICATION OF THERMOPHILIC BIOSURFACTANT PRODUCER

Among 10 isolated potential biosurfactant producers, isolate A was selected for further identification using biochemical tests and 16S rRNA sequence analysis. Biochemical characteristics of the isolate were not shown. The biochemical and physiological features were only suitable as preliminary identification due to their intrinsic limitation [9]. The results of the biochemical tests clearly seemed to coincide with the characteristics belonging to *Anoxybacillus* sp. as reported by Bergey's manual which was further confirmed using BLAST analysis. The *Anoxybacillus* is a genus of thermophilic bacilli comprising 16 species and was first formally described by Pikuta et al. [15]. Pikuta et al. [15] reported that these species are Gram-positive bacteria, motile, producing terminal spore, round shaped colonies and smooth, oxidase-positive and catalase-positive. Most of them are isolated from hot spring and are growing within the temperature of 30–75°C. Production of biosurfactant by *Anoxybacillus* sp. has been rarely studied although it was suspected to be a potential candidate that needs to be explored.

8.3.3   STABILITY TO PH, SALINITY, AND TEMPERATURE

The applications of biosurfactants in several fields are affected by the stability of biosurfactant at different environmental factors such as pH, salinity, and temperature. As indicated in Figure 8.1, the minimum surface tension value was obtained at pH 7 (33.97 mN/m). The surface activity of the sample relatively remained stable between pH 10–12 indicating preference for alkaline conditions.

Figure 8.2 showed that surface tension started to decline at 4% to 6% (w/v) of sodium chloride (NaCl) and the lowest surface tension value was achieved at 6% (w/v) of NaCl which was 18.93 mN/m). According to Bergey's manual, at 5–6% (w/v) NaCl the growth of *Anoxybacillus* sp. is inhibited, but from the result, the surface tension activity was stable within that range of NaCl. Reports on the stability of biosurfactant produced by *Anoxybacillus* sp. are very limited. Even though the bacteria cannot afford to live in a condition that has high salt content, it proves that the extracellular biosurfactant produced by this kind of bacteria is able to withstand in high salt content.

Temperature also has some influences on the surface tension activity. As indicated in Table 8.2, the cultured supernatant was reduced from 50.15 to 35.03–33.97 mN/m. The surface tension did not show any significant loss of activity but the most stable conditions was at temperature 25°C (33.97 mN/m). Therefore, it can be concluded that this biosurfactant maintained its surface tension activity and was unaffected within the temperature range of 4–121°C.

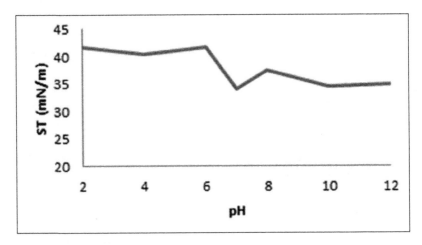

FIGURE 8.1   Effect on surface tension (ST) at different pH range.

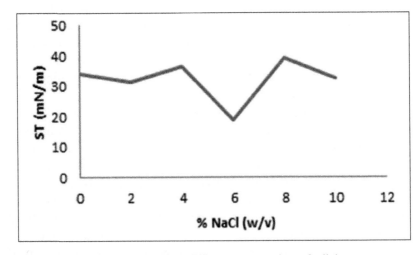

FIGURE 8.2   Surface tension (ST) at different concentrations of salinity.

8.3.4 ANALYSIS OF BIOSURFACTANT USING FT-IR

The IR spectra of purified surfactant were analyzed during 5 days of cultivation (data not shown). The deformation of strong vibration (stretching mode of C–H) at 2926 cm^{-1} and 2853 cm^{-1} confirmed the presence of alkyl groups. In the region of 1460–1465 cm^{-1} and 1365–1380 cm^{-1}, several fractions were observed to reflect toward aliphatic chains (-CH3, -CH2-). It can be clearly observed that the characteristic absorbance bands of peptides at 3405 cm^{-1} (NH-stretching mode) and 1648 cm^{-1} (stretching mode of the CO–N). A broad absorption valley in the range of 3100–3600 cm^{-1} was also present. This region is a typical feature of compound containing carbon and amino groups and is due to the stretching vibrations of C–H and N–H bonds. Absorption valley at 1740 cm^{-1} and 1117 cm^{-1} are stretching vibration of C–O and C=O bonds in carboxyl ester, respectively. Scissoring vibration of CH2 group adjoining a carboxyl ester was also observed at 1370 cm^{-1}.

8.5 CONCLUSION

Isolate A that was isolated from hot spring Sungai Klah, Perak, Malaysia was identified as *Anoxybacillus* sp. and a biosurfactant producer. *Anoxybacillus* sp. is a new genus compared to the well-studied *Geobacillus* sp. or *Bacillus* sp. Application of biosurfactant and biosurfactant-producing bacteria in environmental cleaning is a potential area of more research as revealed from the present study. Due to their biodegrading potential and low hazard to the environment and human health, the applications of these microbes are very promising in environmental and biotechnological applications. The commercial success of such technologies is still limited by their high production cost. Optimized growth conditions using cheap renewable substrates (agro-industrial wastes) and efficient methods for cultivation of microbes could make these technologies more economically feasible.

8.6 ACKNOWLEDGMENT

The author would like to thank Research Management Institute (RMI) Universiti Teknologi Mara (UiTM) for funding this project (600-RMI/

DANA 5/3/RIF (41/2012) and Faculty Applied Science (FSG), UiTM. Special thanks to Assoc. Prof. Dr. Tengku Elida from FSG, Ionic, Color and Coating Laboratory (ICC) for his kindness and help in the surface tension measurements.

KEYWORDS

- **biosurfactant**
- *Anoxybacillus*
- **surface tension**
- **temperature**
- **culture**

REFERENCES

1. Ariji, A. L.; Rahman, A. R. Z. N. R.; Basri, M.; Salleh, B. A. Microbial Surfactant. *Asia Pac. J. Mol. Biol. Biotech.* **2007,** *15*(3), 99–105.
2. Banat, I. M. The Isolation of Thermophilic Biosurfactant Producing *Bacillus* sp. *Biotechnol. Lett.* **1993,** *15*, 591–594.
3. Bodour, A. A.; Maier, R. M. Biosurfactants: Types, Screening Methods and Application. *Encycl. Environ. Microbiol.* **2002,** *2*, 750–769.
4. Bodour, A. A.; Miller-Maier, R. M. Application of Modified Drop-Collapse Technique for Surfactant Quantitation and Screening of Biosurfactant-Producing Microorganisms. *J. Microbiol. Meth.* **1998,** *32*, 273–280.
5. Bosch, M. P.; Robert, M.; Mercade, M. E.; Espuny, M. J.; Parra, J. L.; Guinea, J. Surface-Active Compounds on Microbial Cultures. *Tenside Surfact. Det.* **1998,** *25*, 208–211.
6. Carrillo, P. G.; Mardaraz, C.; Pitta-Alvarez, S. J.; Giulietti, A. M. Isolation and Selection of Biosurfactant-Producing Bacteria. *World J. Microbiol. Biotechnol.* **1996,** *12*, 82–84.
7. De Clerck, E.; Rodriguez-Diaz, M.; Vanhoutte, T.; Heyrman, J.; Logan, N. A.; DeVos, P. Anoxybacillus Contaminans sp. nov. and *Bacillus* gelatini sp. nov., Isolated from Contaminated Gelatin Batches. *Int. J. Syst. Evol. Micr.* **2004,** 941–946.
8. Kadriye, I.; Sabriye, C.; Ali Osman, B. Isolation and Characterization of Xylanolytic New Strains of *Anoxybacillus* from Some Hot Springs in Turkey. *Turk. J. Biol.* **2011,** *35*, 529–542. doi: 10.3906/biy-1003-75

9. Huang, X. F.; Guan, W.; Liu, J.; Lu, L. J.; Xu, J. C.; Zhou, Q. Characterization and Phylogenetic Analysis of Bio Demulsifier Producing Bacteria. *Bioresource Technol.* **2010**, *101*, 317–323.

10. Hussin, N.; Akmar, I.; Asma, B.; Venugopal, L; Yoga L; Sasidharan, S. Identification of Appropriate Sample and Culture Method for Isolation of New Thermophilic Bacteria from Hot Spring. *Afr. J. Microbiol. Res.* **2011**, *5*(3), 217–221.

11. Jaysree, R. C.; Subham, B.; Priyanka P. S; Twinkle, G.; Pragya, A. P.; Yekala, K.; Rajendran, N. Isolation of Biosurfactant Producing Bacteria from Environmental Samples. *Pharmacologyonline* **2011**, *3*, 1427–1433.

12. Makkar, R. S.; Cameotra, S. S. Biosurfactant Production by a Thermophilic *Bacillus* subtilis strain. *J. Ind. Microbiol. Biotechnol,.* **1997**, *18*, 37–42.

13. Mulligan, C. N.; Cooper, D. G.; Neufeld, R. J. Selection of Microbes Producing Biosurfactants in Media without Hydrocarbons. *J. Ferment. Technol.* **1984**, *62*, 311–314.

14. Pakpitcharoena, A.; Potivejkulb, K.; Kanjanavasa, P.; Areekita, S.; Chansiria, K. Biodiversity of Thermotolerant *Bacillus* sp. Producing Biosurfactants, Biocatalysts, and Antimicrobial Agents. *ScienceAsia,* **2008**, *34*, 424–431. DOI: 10.2306/scienceasia1513-1874.2008.34.424.

15. Pikuta, E.; Lysenko, A.; Chuvilskaya, N.; Mendrock, U.; Hippe, H.; Suzina, N.; Nikitin, D.; Osipov, G.; Laurinavichius, K. *Anoxybacillus pushchinensis* gen. nov., sp. nov., a Novel Anaerobic, Alkaliphilic, Moderately Thermophilic Bacterium from Manure, and Description of Anoxybacillus Flavithermus Comb. *nov. Int. J. Syst. Evol. Microbiol.* **2000**, *50*, 2109–2117.

16. Plaza, G. A.; Zjiawiony, I.; Banat, I. M. Use of Different Method for Detection of Thermophlic Biosurfactant Producing Bacteria from Hydrocarbon-Contaiminated and Bioremediation Soil. *J. Petrol Sci. Eng.* **2006**, *50*, 71–77.

17. Poli, A.; Esposito, E.; Lama, L.; Orlando, P.; Nicolaus, G., de Appolonia, F.; Gambacorta, A.; Nicolaus, B. Anoxybacillus Amylolyticus sp. nov., a Thermophilic Amylase Producing Bacterium Isolated from Mount Rittmann (Antarctica). *Syst. Appl. Microbiol.* **2006**, *29*, 300–307.

18. Ron, E. Z.; Rosenberg, E. Natural Roles of Biosurfactants. *Environ. Microbiol.* **2001**, *3*, 229–236.

19. Santos, D. C. S.; Fernandez, G. L.; Alva, R. C. J.; Roque, A. D. R. M. Evaluation of Substrates from Renewable-Resources in Biosurfactants Production by Pseudomonas Strains. *Afr. J. Biotechnol.* **2010**, *9*(35), 5704–5711.

20. Schulz, D.; Passeri, A.; Schmidt, M. Screening for Biosurfactants among Crude Oil Degrading Marine Microorganisms from the North Sea. *Mar. Biosurfactants* **1991**, *46*(3–4), 197–203.

21. Singh, A.; Hamme, V. D. J.; Ward, P. O. Surfactants in Microbiology and Biotechnology: Part 2. Application Aspects. *Biotechnol. Adv.* *25*(1), 99–121.

22. Vardor-Suhan, F.; Kosaric, N. Biosurfactants, 2nd ed. Encyclopedia of Microbiology. *Biosurfactants, 2nd ed. Encycl. Microbiol.* **2000**, *1*, 618–635.

23. Walter, V.; Syldatk, C.; Hausmann, R. Screening Concepts for the Isolation of Biosurfactant Producing Microorganisms. *Landes Bioscience.* Madame Curie Bioscience Database [Internet], 2010; pp 1–13.

24. Wilkesman, J.; Florian, B.; Lililana, K.; Peter, M. Biodegradation of an Emulsified Oil Mixture Employing aVenezuelan Thermophilic *Anoxybacillus* sp. Strain. *FARAUTE Ciens. y Tec.* **2008,** *3*(2), 61–67.

25. Willumsen, P. A.; Karlson, U. Screening of Bacteria Isolated from PAH-Contaminated Soils for Production of Biosurfactants and Bioemulsifiers. *Biodegradation* **1997,** *7*, 415–423.

26. Youssef, N. H.; Duncan, K. E.; Nagle, D. P.; Savager, K. N.; Knapp, R. M.; McInerney, M. J. Comparison of Methods to Detect Biosurfactant Production by Diverse Microorganisms. *J. Microbiol. Meth.* **2004,** *56*, 339–346.

CHAPTER 9

TOXICITY EFFECTS OF DIMETHOATE AND CHLORPYRIFOS ON ESTERASE IN TILAPIA (*OREOCHROMIS NILOTICUS*)

ASYSYUURA ADYTIA PATAR, JESSEY ANGAT,
NORDIANA SUHADA MOHAMAD TAHIRUDDIN,
FARIDA ZURAINA MOHD YUSOF, and
WAN ROZIANOOR MOHD HASSAN*

*Faculty of Applied Sciences, Universiti Teknologi MARA, 40450
Shah Alam, Selangor, Malaysia*

Corresponding author. E-mail: rozianoor@salam.uitm.edu.my

CONTENTS

ABSTRACT

Dimethoate and chlorpyrifos are pesticides frequently used for pest control in agricultural fields, especially in Malaysia. These pesticides may reach the surrounding freshwater bodies by direct application or through draining of agricultural watersheds, overspray, and runoff from crop fields after rain. The mode of action of dimethoate and chlorpyrifos are through inhibition of esterase activity. There were two objectives for this study, firstly to determine the inhibitor concentration to inhibit 50% enzyme activity (IC_{50}) and secondly to determine the constant rate of inhibition (k_i). The esterase was extracted from the muscles of the tilapia (*Oreochromis niloticus*) and was stored at $-80°C$. There were four optimum parameters determined: pH of buffer, substrate concentration, temperature, and incubation time. Total protein and enzyme activity were determined by Lowry and van Leeuwen methods, respectively. This study showed the optimum condition to be at pH 7 of phosphate buffer with 12.0×10^{-4} M of α-naphthyl acetate at 25°C for 30 minutes. The enzyme activity marked as 3.29×10^{-7} μmol/mg/min and IC_{50} for both inhibitors was 1×10^{-4} M. Besides that, the k_i values for dimethoate and chlorpyrifos were marked as 4.796 and 0.834, respectively. Therefore, from the values obtained, dimethoate and chlorpyrifos can be indicated as highly toxic since it can inhibit esterase activity at very low concentration. However, from the k_i values, it was showed that chlorpyrifos was more potent compared with dimethoate as the constant rate value was low enough to indicate the high affinity between the enzyme–inhibitor complexes. Furthermore, the information in this study can be coupled with variety of management strategies available such as integrated pest management system to assure the effective use of these pesticides.

9.1   INTRODUCTION

Nowadays, global population growth coupled with continual technological advancement and increase in the generation of new industrial products, including the manufacture of chemicals such as fertilizers and pesticides, has led to an expansion in the levels of xenobiotic compounds in aquatic ecosystems [11]. Thus, contamination levels in aquatic environments have greatly increased in recent years as a consequence of intense human activities, which in some areas have resulted in a substantial impact [23]. The pollution of rivers and lakes with chemicals of anthropogenic origin may have adverse effects such as the waters become unsuitable for drinking and household purposes, irrigation and fish cultivation, and the animal communities involved may suffer seriously. Massive fish kills are recorded rather frequently and changes in the population of the fauna as a consequence of sublethal effects on ecologically important species have also been described [12].

Insecticides are toxic and were designed to repel or kill unwanted organisms and when used for different purposes they may be brought to water bodies to kill or influence the lives of aquatic organisms [7]. Insecticides' applications are recognized worldwide and compounded by their improper use [22, 26]. Insecticides are used extensively in agriculture and industry because they are easy to apply, cost-effective, and in some situations, their application is only a practical method of control. However, benefits of pesticides are not derived without consequences. They are one of the most potentially harmful chemicals and are released into the environment by direct applications, spraying, atmospheric deposition, and surface runoff. Given the fact that insecticides are not selective and affect nontarget organisms, it is not surprising that a chemical that acts on the insects' different systems will elicit similar effects in higher forms of life [6]. Levels of insecticides in superficial waters generally range far below lethal concentrations for aquatic organisms. However, sublethal adverse effects may result from exposure of aquatic organisms to insecticides at environmentally relevant concentrations [4]. Pesticides released in the environment could be used as a model for the study of ecotoxicology because they contaminate air, land, and water, causing adverse effects on the entire range of bacteria to humans. It is well proven that these chemicals are toxic to aquatic arthropods, bees, and fish [19].

Organophosphates comprise a group of chemical compounds extensively used in farming as insecticides, which cause accidental poisoning in living organisms. The toxicity of these compounds is due to the respiratory and cardiac impairment in consequence of autonomic nervous system disorders. The primary effect of organophosphates on vertebrate and invertebrate organisms is the inhibition of the enzyme acetylcholinesterase (AChE), which is responsible for terminating the transmission of the nerve impulse [16].

Dimethoate is an organophosphate insecticide. It has both direct and systemic action against a broad range of insect pests in various crops and pastures and also in the home garden. As with other organophosphorus chemicals, the mode of action of dimethoate is through inhibition of esterase activity [27]. Chlorpyrifos is a synthetic OP, nonsystematic, broad-spectrum insecticide and acaricide, acting as a cholinesterase inhibitor, with contact, stomach, and respiratory action [10].

The red Nile Tilapia, *Oreochromis niloticus* (Cichlidae) is a model fish largely studied in the ecotoxicology field. Several biochemical responses have been well characterized in this fish in response to exposure to toxic compounds [9, 18, 21]. Tilapia has become one of the most important fish in aquaculture because of its large size, rapid growth, and palatability. It is a species of high economic value and is widely introduced outside its natural range.

9.2 MATERIALS AND METHODS

9.2.1 ENZYME PREPARATION

The tilapia fish was obtained from Prima Puchong. It was placed in the freezer at −4°C before the grinding process. The whole muscles of the fish were separated from the head, bones, and tail. About 10 g of fish muscles were placed in a beaker surrounded with ice cubes and been homogenized (0.1 g/mL) in 10 mL of phosphate buffer. Then, the crude homogenates was centrifuged at 2500 rpm for 40 min.

9.2.2 OPTIMUM PARAMETERS DETERMINATION

The parameters studied were pH, substrate concentration, temperature, and incubation time.

9.2.3 PROTEIN DETERMINATION

The protein content was determined by using Lowry's method [14] and bovine serum albumin (BSA) as a reference and absorbance was read at 750 nm. A graph of optical density (OD) versus BSA concentration was plotted to obtain the standard calibration curve.

9.2.4 ENZYME ACTIVITY DETERMINATION

The specific enzyme activity of esterase was determined by using standard curve of α-naphthol and fast blue salt [24]. The absorbance was read at 405 nm. After that graph of OD versus concentration of α-naphthol was plotted in order to get the total concentration value of α-naphthol (μmol/mL) that was substituted into the formula. From the data that were obtained, the enzyme activity was calculated.

$$\text{Esterase activity} = \frac{\text{Total} \left[\alpha - \text{naphthol} \right] \left(\mu\text{mol} / \text{mL} \right)}{\text{Incubation time} \left(\text{min} \right) \times \text{Total protein} \left(\text{mg} / \text{mL} \right)}$$

9.2.5 I$_{50}$ DETERMINATION

Four samples of blank control, control, blank test, and test were prepared. All samples were incubated for 10 min at 25°C and were measured at absorbance 405 nm. From the absorbance reading, the percentage of inhibition was calculated according to the formula. Then, the I$_{50}$ value was obtained from graph of percentage of inhibition versus negative log of inhibitor concentration.

$$\% \text{ of inhibition} = \frac{\text{Absorbance of control} - \text{Absorbance of test}) \times 100\%.}{\text{Absorbance of control}}$$

9.2.6 K$_i$ DETERMINATION

Inhibitor concentration was assumed as χ M. Enzyme–inhibitor (E–I) complex was prepared by mixing 10 mL of inhibitor and 10 mL of enzyme (esterase). Then, the E–I was allowed to react from 1 to 10 min. The reaction was determined for the three samples that were prepared which were blank, control, and test. All mixtures were incubated for t minutes at 25°C. Then, all samples were added with 0.5 mL of coupling reagent and were incubated again for 10 min at 25°C. Approximately 0.5 mL of 10% acetic acid was added to stop the reaction. Absorbance was read at 405 nm. The step was repeated for other time interval. From the absorbance reading, the percentage of remaining activity was calculated. Then, the graph of percentage of remaining activity versus time (1–10 min) was plotted to determine the gradient of the graph. Thus, the k_i value was calculated by multiplying the gradient of graph by constant rate of enzyme, 0.2303.

$$\text{\% of remaining activity} = \frac{\text{Absorbance of test} \times 100\%}{\text{Absorbance of control}}$$

9.3 RESULTS AND DISCUSSION

9.3.1 OPTIMUM PARAMETERS

The optimum value for each parameter was determined by the highest value of OD reading, which indicated that the reaction occurred at its optimum rate. This study showed the optimum condition for the enzyme activity was at pH 7 of phosphate buffer with α-naphthyl acetate concentration of 12.0×10^{-4} M at temperature 25°C for 30 min of incubation time (Table 9.1).

TABLE 9.1　Optimum Parameters.

Parameters	Optimum Value	Optical Density (OD)
pH	7	2.041
α-Naphthyl acetate (M)	12.0×10^{-4}	1.950
Temperature(°C)	25	2.140
Time (min)	30	2.117

These findings were supported by few previous studies; first, for optimum pH, [28] stated that specific activities increased as the pH increased from 6 to 7 and also stated that esterases exhibit maximum activity at pH 7. For optimum substrate concentration, previous research by Davis and Green [5] stated the specific activity of general esterase toward α-naphthyl acetate increased with increasing substrate concentration. Next, for optimum temperature, the result supported the study conducted by Beauvais et al.[1], which stated optimal reaction rates occurred at 25°C for bluegill (*Lepomis macrochirus*) which is also a freshwater fish. Lastly, for optimum incubation time, previous study by Leite et al.[13] on tadpoles mentioned that it took 30 min for the activity to completely occur.

9.3.2 PROTEIN DETERMINATION

Based on the result, the value obtained was 0.4 mg/mL. Low amount of protein fraction in muscle tissues may have been due to degradation and possible utilization of degraded products for metabolic purposes (Fig. 9.1).

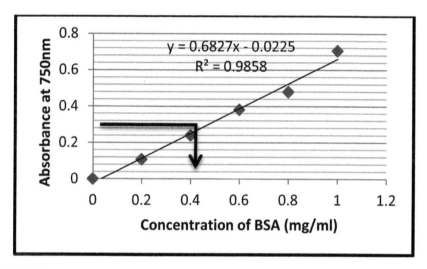

FIGURE 9.1 Graph of BSA standard curve.

9.3.3 ENZYME ACTIVITY DETERMINATION

The enzyme activity of general esterase was determined by using α-naphthol standard curve [24], which the average absorbance reading of α-naphthyl acetate was used to be substituted into the graph (Table 9.2).

TABLE 9.2 Optical Density (OD) Reading of α-Naphthyl Acetate.

Trial	Trial 1	Trial 2	Trial 3	Average
OD	0.206	0.258	0.256	0.240

9.3.4 α-NAPHTHOL STANDARD CURVE DETERMINATION

Referring to the result on the graph of α-naphthol standard curve, it showed positive relationship. The total α-naphthol concentration is 3.95×10^{-6}

µmol/mL. This value is then substituted into the formula of esterase activity, and the total enzyme activity for general esterase was 3.29×10^{-7} µmol/mg/min. This showed the amount of product formed by an enzyme in a given amount of time under given conditions per milligram of total protein (Fig. 9.2).

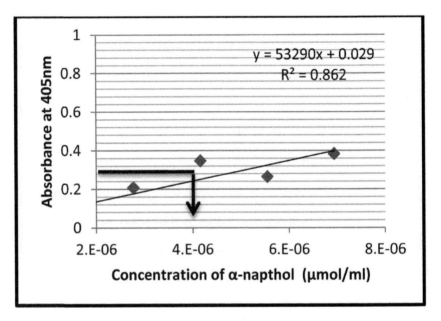

FIGURE 9.2   Graph of α-naphthol standard curve.

9.3.5 I_{50} DETERMINATION

Based on the result obtained, it showed significant effect of dimethoate and chlorpyrifos on the general esterase activity in *O. niloticus*. Figure 9.3 and Figure 9.4 show that only 1×10^{-4} M concentration of dimethoate and chlorpyrifos was required to inhibit 50% of the enzymatic activity in general esterase, which is 3.29×10^{-7} µmol/mg/min. This result proved that even with low concentration, dimethoate and chlorpyrifos are very toxic to the fish.

The degree of inhibition increased with higher exposure of organophosphate [15]. This reaction is a progressive, covalent adduction of the organophosphate to the active site and as such is dependent on concentration.

Thus, the inhibitor blocked the substrate from binding to the active site of the enzyme. However, based on the result obtained, esterase only requires low concentration of OP to inhibit half of the population. This was also supported from previous study, which stated that dimethoate and chlorpyrifos can be rated as highly toxic to nontarget organism [2].

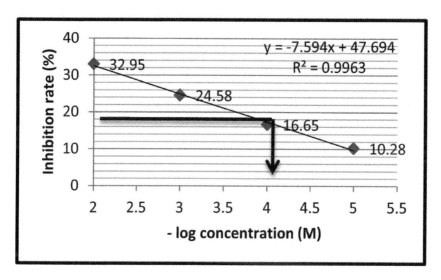

FIGURE 9.3   Graph of 50% inhibition of dimethoate inhibition toward esterase.

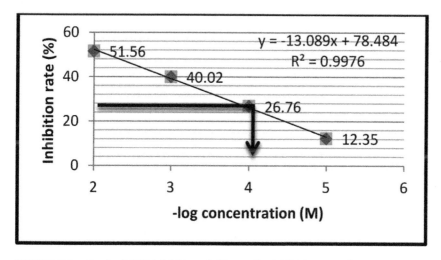

FIGURE 9.4   Graph of 50% inhibition of chlorpyrifos inhibition toward esterase.

At 1×10^{-4} M concentration, dimethoate and chlorpyrifos would act as cholinesterase-competitive inhibitors that interfere with the esterase-binding site in correspondence to their substrates that cause primarily irreversible inhibition of AChE. This would result in the accumulation of neurotransmitter acetylcholine at the synapse [20]. Such accumulation causes hyperactivity, paralysis, and finally death of the organism [8].

Dimethoate and chlorpyrifos affect other esterases such as carboxylesterases as these enzymes are frequently more sensitive to OP inhibition than ChE [25]. These esterases are not directly involved in the acute OP toxicity as AChE is, and this could explain the reduced amount of ecotoxicological investigations with CEs. However, these enzymes play an important role in pesticide detoxification. They hydrolyze efficiently synthetic pyrethroids and carbamates [3].

9.3.6 K_i DETERMINATION

Based on Figure 9.5 and Figure 9.6, the percentage of remaining activity was decreased with the increase in time interval. Referring to the graph, the k_i value, which represented the constant rate of inhibition was calculated

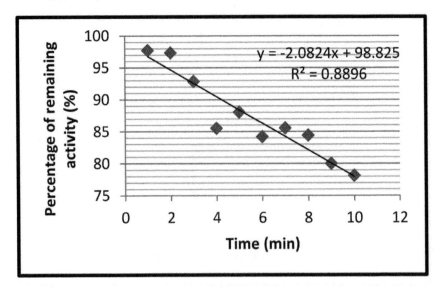

FIGURE 9.5   Graph of percentage of remaining activity over time of dimethoate for *O. niloticus*.

by multiplying the slope of the graph with constant rate of enzyme, 2.303. From the result obtained, it showed that the 50% inhibition of esterase activity for dimethoate and chlorpyrifos occurred at constant rate of 4.796 and 0.834, respectively.

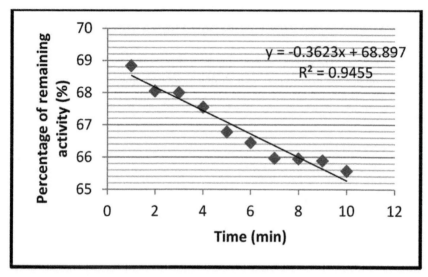

FIGURE 9.6   Graph of percentage of remaining activity over time of chlorpyrifos for *O. niloticus.*

The percentage of remaining activity showed that the amount of specific activity of esterase left as the reaction between the E–I complex for a given period. It h showed that there was a high affinity between the enzyme and inhibitor; thus, less inhibitor is required to inhibit the enzyme [17]. Since the action of the OP that caused the irreversible inhibition of the enzyme, the effects of the reaction would depend on the concentration of the OP as the value of inhibition constant, k_i that the concentration required to produce half-maximum inhibition.

Based on previous study conducted by McHenery et al.[15], the higher the k_i value, the higher the sensitivity of enzymes binds toward inhibitor. It means that the enzyme binds more to inhibitor instead of substrate. In other words, the binding of an inhibitor can stop the substrate from entering the enzyme active site, thus hinder the enzyme from catalyzing its reaction. These inhibitors modify key amino acid residues needed for enzymatic activity.

In addition, McHenery et al.[15] stated that E–Is are often designed to mimic the transition state or intermediate of an enzyme-catalyzed reaction. The enzyme (esterase) has better binding affinity with the inhibitor (dimethoate and chlorpyrifos) instead with the substrate (α-naphthyl acetate) because the inhibitor exploits the transition-state-stabilizing effect of the enzyme.

From the values obtained, dimethoate and chlorpyrifos can be indicated as highly toxic since it can inhibit esterase activity at very low concentration. However, from the k_i values, it showed chlorpyrifos was more potent compared with dimethoate because the constant rate value was low enough to indicate the high affinity between the E–I complexes.

9.4 CONCLUSION

Thus, from this finding and based on several previous studies, it was approved that dimethoate and chlorpyrifos are very toxic pesticides to tilapia (*O. niloticus*) even in very low concentration. Besides, the toxicity of dimethoate and chlorpyrifos was proved to relate with esterase inhibition. Therefore, fish with toxicity effects of dimethoate and chlorpyrifos is not fit for human consumption. Furthermore, precautions in the application of insecticides to protect the life of fish and other aquatic fauna should be implemented. The precaution steps that can be taken are to prevent overuse and leakage of the insecticides into the aquatic environment. Other than that, this information can be coupled with variety of management strategies available such as the integrated pest management system to assure the effective use of these pesticides and to publish a guideline on the safe use of this insecticide in order to avoid harm to living organisms.

9.5 ACKNOWLEDGMENT

This study was supported by Ministry of Higher Education, Faculty of Applied Sciences, Universiti Teknologi MARA, and Research Intensive Faculty Grant (600-RMI/DANA 5/3/RIF (65/2012)).

KEYWORDS

- *Oreochromis niloticus*
- esterase activity
- α-naphthol
- dimethoate
- chlorpyrifos

REFERENCES

1. Beauvais, S. L.; Cole, K. J.; Atchison, G. J.; Coffey, M. Factors Affecting Brain Cholinesterase Activity in Bluegill (*Lepomis macrochirus*). *Water, Air Soil Pollut.* **2002**, *135*, 249–264.

2. Crane, M.; Whitehouse, P.; Comber, S.; Watts, C.; Giddings, J.; Moore, D. R. J.; Gist, E. Evaluation of Probabilistic Risk Assessment of Pesticides in the UK: Chlorpyrifos Use on Top Fruit. *Pest. Manage. Sci.* **2003**, *59*, 512–526.

3. Crow, J. A.; Borazjani, A.; Potter, P. M.; Ross, M. K. Hydrolysis of Pyrethroids by Human and Rat Tissues: Examination of Intestinal, Liver and Serum Carboxylesterases. *Toxicol. Appl. Pharmacol.* **2007**, *221*, 1–12.

4. Das, B. K.; Mukherjee, S. C. Toxicity of Cypermethrin in *Labeo rohita* Fingerlings: Biochemical, Enzymatic and Hematological Consequences. *Comp. Biochem. Physiol.* **2003**, *134*, 109–121.

5. Davis, D. R.; Green, A. L. The Kinetics of Reactivation by Oximes of Cholinesterase Inhibited by Organophosphorus Compounds. *Biochem. J.* **1995**, *63*, 529–535.

6. Dogan D.; Can, C. Hematological, Biochemical, and Behavioral Responses of *Oncorhynchus mykiss* to Dimethoate. *Fish Physiol. Biochem.* **2011**, *37*, 951–958.

7. El-Sayed, Y. S.; Saad, T. T.; El-Bahr, S. M. Acute Intoxication of Deltamethrin in Monosex Nile Tilapia, *Oreochromis niloticus* with Special Reference to the Clinical, Biochemical and Hematological Effects. *Environ. Toxicol. Pharmacol.* **2007**, *24*, 212–217.

8. Fulton. M. H.; Key, P. B. Acetylcholinesterase Inhibition in Estuarine Fish and Invertebrates as Indicator of Organophosphorus Insecticide Exposure and Effects. *Environ. Toxicol. Chem.* **2001**, *20*, 37–45.

9. Gold-Bouchot, G.; Zapta-Perez, O.; Rodriguez-Fuentes, G.; Ceja-Moreno, V.; Rio-Garcia, M. D.; ChzanCocom, E. Biomarkers and Pollutants in the Nile Tilapia, *Oreochromis niloticus*, in Four Lakes from San Miguel, Chiapa, Mexico. *Int. J. Environ. Pollut.* **2006**, *26*(1/2/3), 130–141.

10. Halappa, R.; David, M. Behavioural Responses of the Freshwater Fish, *Cyprinus carpio* (Linnaeus) Following Sublethal Exposure to Chlorpyrifos. *Turkish J. Fish. Aquat. Sci.* **2009**, *9*, 233–238.

11. Jesus, T. B.; Carvalho, C. E. V. Utilização de biomarcadores em peixes como ferramenta para avaliação de contaminação ambiental por mercúrio (Hg). *Oecologia brasiliensis* **2008**, *12*, 7.

12. Koprucu, K.; Aydin, R. The Toxic Effects of Pyrethroiddeltamethrin on the Common Carp (Cyprinus carpio L.) Embryos and Larvae. *Pestic. Biochem. Physiol.* **2004**, *80*, 47–53.

13. Leite, P. Z.; Margarido, T. C. S.; Lima, D.; Rosa-Feres, D. C.; Almeida, E. A. Esterase Inhibition in Tadpoles of *Scinax fuscovarius* (Anura, Hylidae) as a Biomarker for Exposure to Organophosphate Pesticides. *Environ. Sci. Pollut. Res.* **2010**, *17*, 1411–1421.

14. Lowry, O. H.; Rosebrough, N. J.; Farr, A. L.; Randall, R. J. Protein Measurement with Folin Phenol Reagent. *J. Biol. Chem.* **1951**, *193*, 265.

15. McHenery, J. G.; Linley-Adams, G. E.; Moore, D. C.; Rodger, G. K.; Davies, I. M. Experimental and Field Studies of the Effects of Dichlorvos Exposure on Acetylcholinesterase Activity in the Gills of the Mussel, *Mytilus edulis L. Aquat. Toxicol.* **1997**, *38*, 125–143.

16. Peña-Llopis, S.; Ferrando, M. D.; Peña, J. Fish Tolerance to Organophosphate-induced Oxidative Stress is Dependent on the Glutathione Metabolism and Enhanced by N-acetylcysteine. *Aquat. Toxicol.* **2003**, *65*, 337–360.

17. Robert, A. C. *Evaluation of Enzyme Inhibitors in Drug Discovery*. John Wiley & Sons, Inc.: New Jersey, USA, 2005; pp 64–86.

18. Rodriguez-Fuentes, G.; Gold-Bouchot, G. Characterization of Cholinesterase Activity from Different Tissues of Nile Tilapia (*Oreochromis niloticus*). *Mar. Environ. Res.* **2004**, *58*, 505–509.

19. Santos, M.; Reyes, F.; Areas, M. Diversity and Spatio-temporal Distribution of Frogs in Regions with Pronounced Dry Season in Southeastern Brazil. *Zool. Ser.* **2007**, *18*, 3.

20. Taylor, P. Anticholinesterase Agents. In *The Pharmacological Basis of Therapeutics*; Gilman, A. G., Wall, T., Nies, A., Taylor, P. Eds.; Pergamon Press: New York, USA. 1990, pp 131–149.

21. Tridico, C. P.; Rodrigues, A. C. F.; Nogueira, L.; Silva, D. C.; Moreira, A. B.; Almeida, E. A. Biochemical Biomarkers in *Oreochromis niloticus* Exposed to Mixtures of Benzoapyrene and Diazonin. *Ecotoxicol. Environ. Saf.* **2010**, *73*, 858–863.

22. Tsuda, T.; Aoki, S.; Inoue, T.; Kojima, M. Accumulation and Excretion of Diazinon, Fenthion and Fenitrothion by Killifish: Comparison of Individual and Mixed Pesticides. *Water Res.* **1995**, *29*, 455–458.

23. van der Oost, R.; Beyer, J.; Vermeulan, N. P. E. Fish Bioaccumulation and Biomarkers in Environment Risk Assessment a Review. *Environ. Toxicol. Pharmacol.* **2003**, *13*, 57–149.

24. van Leeuwen, T.; Tirry, L.; Nauen, R. Complete Maternal Inheritance of Bifenazate Resistance in *Tetranychus urticate* Koch (Acari: Tetranychidea) and Its Implications for Mode of Action Considerations. *Insect Biochem. Mol. Biol.* **2006**, *36*, 869–877.

25. Wheelock, C. E.; Phillips, B. M.; Anderson, B. S.; Miller, J. L.; Miller, M. J.; Hammock, B. D. Applications of Carboxylesterase Activity in Environmental Monitoring and Toxicity Identification Evaluations (TIEs). *Rev. Environ. Contam. Toxicol.* **2008**, *195*, 117–178.

26. Wilson, C.; Tisdell, C. Why Farmers Continue to Use Pesticides Despite Environmental, Health and Sustainability Costs. *Ecolog. Econom.* **2001**, *39*, 449–462.
27. World Health Organization, WHO. Dimethoate. WHO Specifications and Evaluation for Public Health Pesticides, 2006, 1–31.
28. Xiaodun, H. A Continuous Spectrophotometric Assay for the Determination of Diamondback Moth Esterase Activity. *Arch. Insect Biochem. Physiol.* **2003**, *54*, 68–76.

CHAPTER 10

PRELIMINARY STUDY OF HUMAN EARPRINTS: METHOD DEVELOPMENT ON SAMPLING AND ENHANCEMENT TECHNIQUE OF EARPRINTS ON A NON-POROUS SURFACE

NURULFARHANA ZULKIFLI and RUMIZA ABD RASHID*

Faculty of Applied Sciences, Universiti Teknologi MARA, 40450 Shah Alam, Selangor, Malaysia

Corresponding author. E-mail: rumiza.rashid@gmail.com

CONTENTS

ABSTRACT

There are several factors that need to be studied before earprints can be considered to be one of the forensic tools in personal identification. Earprints that are discovered from the scene can be compared and matched with the earprints of suspect. An earprint is a two-dimensional reproduction of the parts of the external ear that touch or make contact with different surfaces and produce print. It is typically found in the burglary case or at the time of entering or breaking house or shop. Apart from it, earprints can be used at least for the purpose of elimination. In order to obtain a good print for comparison in investigations, the enhancement methods should have been established before this evidence is applied in investigation. In this study, five subjects were asked to provide their earprints by several different procedures and conditions to see which one is suitable to develop the earprint best. The results showed the influence of pressure distortion, sampling condition, and also grooming techniques toward earprints. Features that frequently appeared in earprints are helix, anti-helix, tragus, and anti-tragus.

10.1   INTRODUCTION

An earprint is a two-dimensional reproduction of parts of external ear that touch surfaces [1]. Oils and waxes that are naturally present on the ear are used to produce prints on surfaces. Earprint and ear pattern may be used to accept or reject the corroboration of two different types of evidence to confirm the presence of a suspect at the crime scene. Previous studies have shown that earprint may be used as one of the supportive tools in forensic discrimination and identification at a crime scene [2]. Information from earprint and ear pattern is more dependable as evidence in court because it cannot be tampered with or accidentally introduced to crime scene. It is almost tamperproof because prints are usually created when someone is intentionally listening at a door or window [2]. Earprints are usually left against the wall, door, mirror, or other hard surfaces during struggling to hear whether the house is occupied or not or when a body is being positioned or moved [3]. It is also an attractive alternative tool because it is cheaper than DNA proof and it is helpful in case where there is no fingerprint or DNA available at the crime scene. Furthermore, not all types of evidence will be equally present in all crime scenes. In forensics, evidential value of earprints is determined by inter- and intraindividual variation. Knowledge about how to select and use earprint features and also the factors that determine variation in earprints must be studied in order to strengthen the scientific basis for earprint individualization [2]. The source of variation should be explored at various stages and it must be identified and considered when conducting the research.

The factors that need to be considered in analysis and comparison of earprint evidence are pressure distortion factors (force applied by the ear on the surface), amount of wax present on external ear when they imprinted on the surfaces, and the best method of enhancement from different surfaces. To obtain good earprint evidence from crime scene for the purpose of comparison, there is an urgent need to establish the enhancement methods for these particular samples. Enhancement method of fingerprint has already been developed by many researchers a few decades ago [4–6]. However, since fingerprint comprises high amino acid than wax or fatty acid when compared with earprint [7], enhancement method of both types of prints may have differences in terms of reactions between chemicals applied with compositions of the prints.

One of the common development methods for prints is a powdering method. It is suitable for enhancing prints on relatively clean and smooth nonporous surfaces. Visualization of latent print with powder, or dusting, involves the application of finely divided particles that stick physically to the aqueous and oily components in latent print residue on nonporous surfaces [8]. Normally, powders are applied to nonporous surfaces with a soft brush. Conventional brushes are usually made with animal hair, fiberglass filaments, or sometimes feathers. Powdering is not recommended for porous or highly absorbent surfaces such as uncoated paper or raw wood because other chemical treatments outperform powder on these surfaces [7]. The softness of the bristles is particularly important to prevent damage to fragile latent print residue. Latent prints with a high moisture or oil content are easily damaged by a brush that is too stiff or is used with excessive force.

This study was performed in order to obtain the best enhancement method of earprint deposited on nonporous surface. PVC wallpaper was chosen as the nonporous surface since this type of wallpaper is the most popular wall decoration in domestic and business outlets. PVC wallpaper is a type of paper that have been coated or treated with polyvinyl chloride and it has been used widely because it is easy to clean and unaffected by long-term exposure to grease and humidity.

10.2   MATERIALS AND METHODS

10.2.1   COLLECTION OF EARPRINT SAMPLES

Samples were collected from five different individuals, with four different procedures and conditions of collection, respectively. All samples were taken on PVC-type wallpaper as the sampling medium to represent the nonporous surface. Factors such as pressures, sampling condition, and grooming technique were considered during sample collection.

10.2.1.1   UNCONTROLLED CONDITION

The uncontrolled condition means earprint was collected in an uncontrolled environment (no apparatus involved). It was divided into two groups which were no grooming technique and with grooming technique.

10.2.1.1.1   No Grooming Technique

No grooming technique means that collections of earprints were made without cleaning the ear first. The latent prints produced will be dependent on oils and waxes present on respondent's ear at the time of collection. Collections of earprints were made by only one examiner to avoid different pressures applied by different persons. In this technique, glass plate was used to support the PVC wallpaper when collecting the samples. The size of glass plate and PVC wallpaper were 10 × 10 cm. The PVC wallpaper was attached properly on top of a glass plate. Subject was asked to sit in a relaxed and straight position on a chair to deposit his or her earprints. Sampling medium was put by side of ear and examiner pushed the sampling medium to subject's ear. Three pressures were applied as variable factors which were soft, medium, and hard pressure. The pressures were determined by examiner with the help of subject.

10.2.1.1.2   With Grooming Technique

In this procedure, grooming technique was applied on the ear and print was collected in an uncontrolled condition. Uncontrolled condition means prints that were collected manually by examiner and there is no apparatus involved. Grooming technique is one of the techniques that substituted the natural oils and waxes present on ear with body lotion that has similar oil and wax content. This technique was performed to avoid the variety of amount of oils and waxes present on subject's ear at the time of collection.

For grooming technique, one type of body lotion was used since the ingredients are similar with original contents of oils and waxes that naturally occurred on human ear. It contains squalene, free fatty acids, and glycerides. Firstly, ear was cleaned using wet tissue to remove any oils and dirt present on ear area. Then ear was air-dried for about 30 s to 1 min. Cotton bud was used to take one drop of lotion and it was swiped on external and internal area of ears. Ears were air-dried for a few minutes to let the lotion dry. The optimized time for drying the lotion was about 6–7 min. After the lotion was completely dried, earprint samples were collected.

Method used in this sampling procedure was similar to the first technique in which glass plate was used to support the PVC wallpaper when collecting the sample. Three pressures were applied using the same procedure as in no grooming technique.

10.2.1.2   CONTROLLED CONDITION

Sampling of controlled condition was done using a sampling box that has been constructed to suit latent earprint sample collection procedure (Fig. 10.1). It was divided into two groups which were no grooming technique and with grooming technique.

10.2.1.2.1   No Grooming Technique

The box (Fig. 10.1) consists of detachable sampling medium which is 10 × 10 cm glass plate that was placed in front of the box, meanwhile the other side is accessible for capturing image and setting up sources of listening. During sample collection, subjects were asked to listen to sound produced from a radio put inside the box, as a simulation of natural listening process or overhearing of sounds against the door or wall at the time of entering or breaking a house or building. Three levels of sounds were used and were set by controlling the radio volume between 5, 10, and 15 that was from slow to loud. The force applied in this procedure was from natural force that came from subject itself based on volume that was set up from slow to loud. The sampling box was put on an adjustable height stool, and sampling glass plate was at the same level with their ears. On this section, no grooming technique was applied and latent earprint produced was dependent on natural oil and waxes on subject's ear.

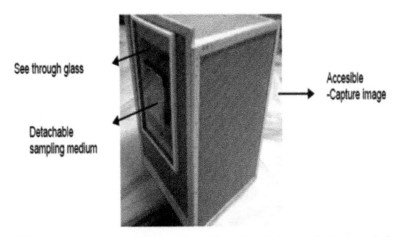

FIGURE 10.1   Subjects were asked to sit upright with their head in the horizontal plane, in a way that was comfortable to them.

10.2.1.2.2   *With Grooming Technique*

Earprints were taken in controlled condition. Grooming technique was applied in this section in which ear was cleaned first then applied with a lotion which has its content similar with ear's oils and waxes. The ear was thoroughly cleaned with wet tissue, and then air-dried for about 30 s to 1 min. After that, using a cotton bud, one drop of body lotion was applied to the external and also internal area of ear. Lotion applied on the ear was air-dried for 6–7 min so that it dried completely.

10.2.2   *ENHANCEMENT OF LATENT PRINT*

All earprint samples collected were immediately enhanced with powdering method with HI-FI Volcano black powder (Sirchie). This method has been established and compared with other chemical enhancement techniques such as ninhydrin, DFO (1,8-diazafluoren-9-one) spray method, and phenolphthalein before sampling from five respondents was done. From four enhancement techniques applied on latent earprint samples, powdering technique is the best to be used on this type of samples. Earprint features were identified after enhancement procedures (Fig. 10.2). In enhancement procedure with powdering technique, latent earprints were dusted with black powder using feather brush to enhance and develop the earprints. The developed earprints were then lifted using lifting tape to preserve the prints. The behavior or pattern of each earprint according to the condition and pressure applied was observed.

10.3   RESULTS AND DISCUSSION

From the early analysis of earprint using different types of enhancement techniques, we found that earprint on nonporous surface is the best to be enhanced using black powder (Fig. 10.2). Apart from earprint, the development of latent fingerprints using powder is one of the earliest known techniques dating back to the 19th century [10]. Dusting technique using black powder is also known to be among the best methods for fingerprint development. The constituents of latent print deposit assist the adherence

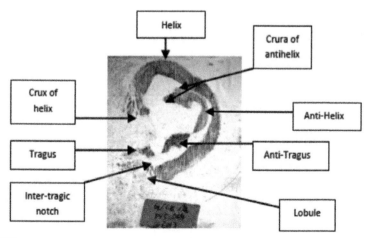

FIGURE 10.2   Earprint enhanced by black powder. Nine earprint features were used to identify earprint samples based on ear morphology (Reprinted from Kieckhoefer, H.; Ingleby, M.; Lucas, G. Monitoring the Physical Formation of Earprints: Optical and Pressure Mapping. *Measurement* **2006**, *39*, 918–935. © 2006 with permission from Elsevier.)

of powder particles thus enhance the impressions to be visible [11]. Developed impressions are then preserved by lifting or photography techniques.

For earprint samples, sampling has been done at several conditions to obtain the best condition for earprint sampling and shall be applied in further research on anthropometry of earprint analysis. Parameters that have been taken into consideration during sampling were grooming technique and pressure applied. Earprint samples were deposited on PVC wallpaper attached to glass plate, either directly collected by pressing the plate on ear, or another way was using an apparatus which required respondents to listen to sound produced from the apparatus.

Based on Figures 10.3 and 10.4, features of earprint that were developed with grooming technique can be clearly seen compared with those

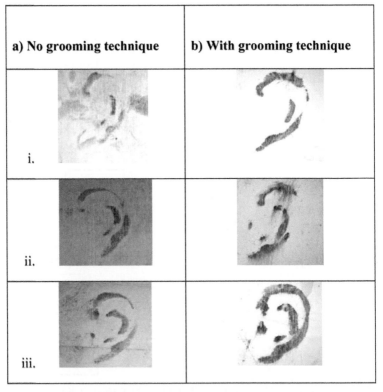

FIGURE 10.3   The best enhanced earprint obtained from one respondent which has the most earprint features among others. Samples were taken on uncontrolled condition (a) with no grooming technique and (b) with grooming technique applied, on different levels of pressure (i) soft, (ii) medium, (iii) hard.

with no grooming technique regardless of controlled and uncontrolled condition. Grooming technique is important in this method because we can standardize the collection of earprint samples from each respondent. Furthermore, with grooming technique, the reproducibility and repeatability of earprint may be accepted for analysis. It is because earprint does not depend on ear wax from respondent which may vary between respondents. From the result also, we can see that earprint with grooming technique on controlled condition showed more features especially tragus and crux of helix which cannot be seen on all pressures with grooming technique on uncontrolled condition.

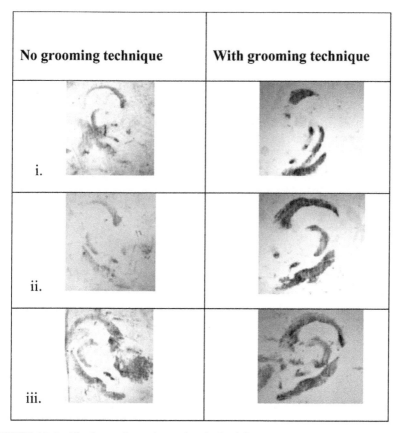

| No grooming technique | With grooming technique |

FIGURE 10.4 The best enhanced earprint obtained from one respondent which has the most earprint features among others. Samples were taken on controlled condition (a) with no grooming technique and (b) with grooming technique applied, on different levels of pressure (i) soft, (ii) medium, (iii) hard.

Detailed analysis on the earprint features was done on all samples and the occurrence of the features was identified. Tables 10.1, 10.2, and 10.3 analyzed the frequency of earprint features appeared in five respondents (n = 5) obtained from enhanced earprint pattern with soft, medium and hard pressure. From the analysis, some common features, especially the outer part such as helix, antihelix, auricular tubercle, antitragus, and earlobe, were clearly seen from most of the samples, from different levels of pressure. Inner part of the ear required extra pressure and grooming technique in order for them to appear. Those features such as tragus, crux of helix, intertragic notch could only be seen from medium and hard pressure samples. It may be because tragus position is at the inner part of the ear, so slightly more pressure is needed for its appearance. Other ear features that were observed in this study such as crura of antihelix only can be seen for hard pressure. For soft pressure applied, only some of the common features such as helix, antihelix, antitragus, and earlobe (lobule) can be seen clearly [2, 12]. Crura of antihelix were absent for soft pressure.

TABLE 10.1   Frequency of Earprint Features in Five Respondents (n = 5) Obtained from Enhanced Earprint Pattern; Soft Pressure Applied.

Features	Uncontrolled Condition		Controlled Condition	
	No Grooming	With Grooming	No Grooming	With Grooming
Helix	5	5	5	5
Antihelix	5	5	4	5
Crux of helix	0	0	1	0
Crura of antihelix	0	0	0	0
Tragus	0	1	0	0
Antitragus	3	3	3	4
Intertragic notch	1	1	1	0
Lobule	3	3	2	1

As in medical term, crura of antihelix are two ridges, that is superior and inferior which are bounding triangular fossa at the upper part of antihelix. As in Figures 10.3 and 10.4 (iii), which is hard pressure applied, most of the features can be seen clearly but their prints were not as neat as medium pressure. With hard pressure application, most of the prints appeared bigger than the real size of ear.

TABLE 10.2 Frequency of Earprint Features in Five Respondents ($n = 5$) Obtained from Enhanced Earprint Pattern; Medium Pressure Applied.

Features	Uncontrolled Condition		Controlled Condition	
	No Grooming	With Grooming	No Grooming	With Grooming
Helix	5	5	5	5
Antihelix	5	5	5	5
Crux of helix	1	1	2	3
Crura of antihelix	0	0	0	0
Tragus	0	2	4	4
Antitragus	3	3	5	4
Intertragic notch	0	2	3	4
Lobule	2	2	3	4

TABLE 10.3 Frequency of Earprint Features in Five Respondents ($n = 5$) Obtained from Enhanced Earprint Pattern; Hard Pressure Applied.

Features	Uncontrolled Condition		Controlled Condition	
	No Grooming	With Grooming	No Grooming	With Grooming
Helix	5	5	5	5
Antihelix	5	5	5	5
Crux of helix	4	3	3	4
Crura of antihelix	3	3	3	4
Tragus	3	4	4	4
Antitragus	5	5	4	5
Intertragic notch	4	5	4	4
Lobule	3	4	4	4

From this study, it is suggested that if earprints were deposited and developed properly, they may be used and help at least in the identification and elimination of individuals. Even though some ears will not simply produce prints with sufficient details that can offer high evidential value, it may still be useful for forensic investigations as they may exclude potential suspect and raise attention to others. Based on these class characteristics or features, it may provide lead in an investigation and individualization of criminal may be established [12]. Further analyses on earprints need to be done by increasing sample size and other means of condition or development of technique.

10.4 CONCLUSION

From this study, it was found that earprint samples obtained from controlled conditions using sampling apparatus, with grooming technique applied prior to sampling and also medium pressure administered on surface studied were the most optimum conditions for earprint sampling. Earprint samples were best to be enhanced using HI-FI Volcano black powder.

ACKNOWLEDGMENT

We would like to thank all subjects that volunteered to lend us their ears for this method development and spend their time for this project. We would also like to thank MOSTI (Ministry of Science, Technology and Innovation) for the research fund, 06-01-01-SF0428.

KEYWORDS

- earprint
- ear pattern
- forensic anthropology
- nonporous surface
- latent print

REFERENCES

1. Meijerman, L.; Thean, A.; Van der Lugt, C.; Van Munster, R.; Van Antwerpen, G.; Maat, G. J. R. Individualization of Earprints: Variation in Prints of Monozygotic Twins. *Forensic Sci. Med. Pathol.* **2005,** *2* (1), 39.
2. Meijerman, L.; Thean, A.; Maat, G. J. R. Earprint in Forensic Investigation. *Forensic Sci. Med. Pathol.* **2005,** *1* (4), 247–256.
3. Meijerman, L.; Sholl, S.; Conti, F. D. Exploratory Study on Classification and Individualisation of Ear Prints. *Forensic Sci. Int.* **2004,** *140,* 91–99.
4. Christophe, C.; Evett, W. E. Probabilistic Approach to Fingerprint Evidence. *J. Forensic Sci.* **2001,** *51,* 101–122.

5. Trozzi, T.; Schwartz, R.; Hollars, M. *Processing Guide for Developing Latent Prints*; Federal Bureau of Investigations, Laboratory Division, U.S. Department of Justice, U.S. Government Printing Office: Washington, DC, 2000.

6. Kent, T. Latent Fingerprint and Their Detection. *J. Forensic Sci. Soc.* **1981,** *21* (1), 15–22.

7. Yamashita, B.; French, M. Latent Print Development, Chapter 7. In *The Fingerprint Sourcebook*. National Criminal Justice Reference Service, **2010;** pp 7–68.

8. Sodhi, G. S.; Kaur, J. Powder Method for Detecting Latent Fingerprint: A Review. *Forensic Sci. Int.* **2001,** *120*, 172–176.

9. Kieckhoefer, H.; Ingleby, M.; Lucas, G. Monitoring the Physical Formation of Earprints: Optical and Pressure Mapping. *Measurement* **2006,** *39*, 918–935.

10. Lee, H.; Gaensslen, R., Eds.; *Advances in Fingerprint Technology*, 2nd Ed.; CRC Press: Boca Raton, FL, 2001; p 108.

11. Henry, J. S.; Amaliya, T. K. Fingerprint Powders: Aerosolized Application Revisited. *J. Forensic Identif.* **2012,** *62* (2), 109.

12. Dhanda, V.; Badhan, J. S.; Garg, R. K. Studies of the Development of Latent Ear Prints and Their Significance in Personal Identification. *Probl. Forensic Sci.* **2011,** *LXXXVIII*, 285–295.

CHAPTER 11

INFLUENCE OF DICLOFENAC ON DEVELOPMENT RATES OF FORENSIC BLOWFLY *CHRYSOMYA MEGACEPHALA*

SITI AISYAH BINTI SHAMSUDDIN*,
NURUL FARHANA BINTI ZULKIFLI,
FARIDA ZURAINA BINTI MOHD YUSOF,
KHAIRULMAZIDAH BINTI MOHAMED, and
RUMIZA BINTI ABD RASHID

Faculty of Applied Sciences, Universiti Teknologi MARA, Shah Alam, Selangor, Malaysia.

Corresponding author. E-mail: siti.aisyah1007@gmail.com

CONTENTS

ABSTRACT

The use of entomological evidence on estimating the post mortem interval (PMI) has been established especially in advanced decay stage. Blowfly *Chrysomya megacephala* is among the most abundant carrion flies applied in PMI estimation. However, the estimation may differ with the presence of drug or toxin since those toxic agents disrupt the normal development of blowflies. Diclofenac is analgesic nonsteroidal anti-inflammatory drug (NSAID) which has been reported to be abused and encountered in drug poisoning cases. The aims of this study were to determine the development rate of *C. megacephala* with the influence of diclofenac and to explore the potential of blowfly samples in determination of diclofenac. Five cow livers homogenized with different increment doses of diclofenac (0 mg/g, 0.5 mg/g, 1.5 mg/g, 4.5 mg/g, and 7.0 mg/g) were introduced to newly emerged *C. megacephala* larvae. Development rate of blowfly was monitored every 6 hours by the measurement of length and width. Research found that larvae contained diclofenac at all doses were shorter compared with the control. Post hoc ANOVA revealed the mean difference was significance ($p < 0.05$) between control and diclofenac groups at the early stage of development (12 hours to 78 hours). Control group completed its development within 198 hours, and it was shorter compared with the highest dose of diclofenac (7.0 mg/g) which was 222 hours. Chemical analysis of *C. megacephala* samples identified the presence of diclofenac. This research revealed the presence of diclofenac delayed *C. megacephala* development for up to 24 hours, thus may give an impact in estimation of PMI. Larvae of *C. megacephala* may have the potential to be used in toxicological analysis when the biological samples are absent at the crime scene due to high decomposition.

11.1   INTRODUCTION

The necrophagous insects are the most important evidence when dealing with decomposed body. Forensic entomology is a study of insects and other arthropods which aid in medico-criminal investigation [1, 2]. The entomological evidence especially flies was used in estimating the time since death or post mortem interval (PMI) either by development rate or successional pattern of carrions insect. Moreover, the carrion insects were also valuable in determining whether the corpse has been removed from origin, providing DNA sample which link the victim with the suspect as well as detection of drugs and toxins consumed by the deceased before death [3, 4].

Commonly, forensic pathologist would be able to estimate the time interval of death by examining the early post-mortem changes such as *livor mortis, rigor mortis,* and *algor mortis.* However, the validity of PMI estimation might be difficult to examine in a period of more than 72h after death due to the beginning of decomposition process of the dead body [5]. Because of that, forensic investigator might use other alternative tool in estimation of PMI by using entomological sample, either by examining the development rates of necrophagous insect or by observation of arthropods successional pattern [6, 7].

Entomotoxicology is an advanced tool in forensic investigation which can provide valuable information in lethal intoxication cases using entomological specimens. It could assist in providing reliable data on the determination of toxic agents such as pesticide, metals, or drugs. Furthermore, biological samples such as blood, urine, and body tissues would be difficult to collect for toxicological analysis during decomposition process. This might be due to the lack or absence of biological samples left on decomposed body since the flies larvae would actively feed on corpse. Moreover, the biological samples collected from corpse would be contaminated with the decomposition fluids resulting from putrefaction process. Thus, there is a need to overcome the problem with other alternative specimens by using entomological evidence, especially fly larvae and beetles. The necrophagous samples feeding on corpse have a potential to be applied as alternative evidence instead of rotten tissue present on the cadaver [8, 9]. In contrast, the estimation of PMI might deviate in the presence of drug or toxin by disturbing the development of blowfly [4–10].

Blowfly (Calliphoridae) and flesh fly (Sarcophagidae) from order Diptera are the first carrion species to arrive on a corpse minutes after death. These carrion flies were attracted to the odour of blood and flesh tissue from the corpse. *Chrysomya megacephala* is among the blowfly species which are predominant on corpse and this species was reported to be found on the decomposed body collected in Thailand and Malaysia [11, 12]. Previous studies done on arthropods successions along decomposition process using animal models in Malaysia have identified the dominancy of *C. megacephala* on cadaver in both indoor [13, 14] and outdoor sites [15, 16]. Because of that, the development of *C. megacephala* was used in this study in order to estimate the PMI.

Diclofenac [(2,6-dichlorophenyl)amino phenyl acetate] known by the trade name Voltaren is one of the nonsteroidal anti-inflammatory drugs (NSAID). This type of drug was among the medicinal drugs which were widely used to treat several rheumatoid disorders, including osteoarthritis, rheumatoid arthritis, acute muscle pain, and other inflammatory conditions [17]. NSAID was reported as the most popular over-the-counter analgesic drug over the world and tend to be abused since it was easily accessed. Several case studies have indicated that NSAID might be involved in drug poisoning due to the adverse effect on gastrointestinal tract ulcer and hepatic injury [18–21]. This drug was also reported in lethal poisoning of livestock in South Asia, Europe, and Africa since diclofenac is widely used by veterinary to treat the animals [22, 23]. Study done by Green et al. [24] found that the rapid decline of vultures of *Gyps* species in India was caused by diclofenac poisoning. Post-mortem from diclofenac-carcass finding has showed the kidney failure due to the toxic effect after overdose exposure to the drug.

The presence of chemical substance in corpse tissue ingested by carrion-feeding arthropods could affect the development rate of insects as well as alter the PMI estimation. This is because the growth rate of necrophagous insect was primarily used by entomologist in estimation of the PMI [9]. The influence of toxicant on necrophagous insects especially blowfly was the major issue in forensic entomology [25], hence would be of concern in this study. Furthermore, qualitative analysis of entomological evidence in determining drugs or toxins could provide such clue in forensic investigation. Because of the study focused on NSAID in entomotoxicology is relatively rare, thus more effort would be required to reveal the finding which could be used as reference by forensic entomologist.

The aims of this study were to determine the development rate of blowfly *C. megacephala* under the influence of diclofenac and to analyze the potential use of blowfly sample in determination of diclofenac using high performance liquid chromatography.

11.2   MATERIALS AND METHODS

11.2.1   *EXPOSURE OF CHRYSOMYA MEGACEPHALA LARVAE TO ARTIFICIAL DIETS CONTAINING DICLOFENAC*

The method was based on Rashid et al. [26] with slight modification. Six cow livers weighing 200g each were used as immature blowfly food. Minced livers were spiked with four doses of diclofenac sodium (Voltaren) between 0.5 and 7 mg/g. The doses were based on lethal dose (LD) of diclofenac sodium in *Gyps bengalensis* [22] which were ½ LD (0.5 mg/g), LD (1.5 mg/g), 3 LD (4.5 mg/g), and 4 LD (7.0 mg/g). One of the untreated minced livers was used as a control. Two hundred newly emerged first instar larvae of blowfly *C. megacephala* from third generation of laboratory colonies were introduced to each of the diets. The larvae were maintained in rectangular transparent containers (18 cm ×12 cm ×8 cm) under controlled room temperature of $27 \pm 1°C$, relative humidity $70 \pm 10\%$ and photoperiod of 12 light:12 dark. Development rates of larvae were monitored every 6h by the measurement of body length and width [27]. Ten larvae from each treatment were preserved using the method described by Adam and Hall [28]. The larvae were killed in hot water at 80°C for 10 min to obtain accurate length and then preserved in 70% ethanol prior to chemical analysis.

11.2.2 SAMPLE PRE-TREATMENT FOR CHEMICAL ANALYSIS

Blowfly sample weighed about 1.0 g and was washed with distilled water before ground using mortar and pestle. Then, 4 mL of acetonitrile was added and homogenized in test tube using vortex about 30s. The mixed sample was centrifuged at 3000 rpm for 10 min. The supernatant was separated and filtered using polytetrafluoroethylene (PTFE) syringe filter 0.45 µm. The aliquot was evaporated with nitrogen gas. Then, the aliquot was reconstituted with 1 mL acetonitrile prior to analysis.

11.2.3 PREPARATION OF STANDARD FOR CALIBRATION CURVE

Six calibration standards at concentrations of 15, 60, 150, 300, 500, and 1000µg/L were prepared by appropriate working solution from stock solution of 100 mg/L diclofenac.

11.2.4 HIGH PERFORMANCE LIQUID CHROMATOGRAPHY ANALYSIS

Chemical analysis was performed using HPLC (Agilent Technologies, USA) coupled with UV/visible detector. The detection and quantification were achieved using RP- C8 column (250 mm × 4.6 mm, 5 m) with mobile phase consisting of acetonitrile and phosphate buffer (10 mM, pH3) (60:40). The analysis was performed at a flow rate of 1.0 mL/min and injection size of 20 µL. The detection of diclofenac was monitored with UV detector set at a wavelength of 280 nm.

11.2.5   STATISTICAL ANALYSIS

The data of blowfly development collected from study were analyzed using SPSS version 21.0. Statistical analysis of variance (ANOVA) was performed to determine the mean differences between control larvae with the treated groups.

11.3   RESULTS AND DISCUSSION

11.3.1   DEVELOPMENT RATE OF CHRYSOMYA MEGACEPHALA

The present study found that there was a decreasing trend on larvae development as the concentration of treatment increased. As can be seen from Figure 1, the development rate of larvae fed on a diet containing diclofenac at all doses was less compared with control. Development rate of blowfly observed was found to exhibit retardation in the presence of diclofenac. Post- hoc analysis of ANOVA revealed the mean differences were significant ($p < 0.05$) between control and diclofenac groups only at the early stage of development which was between 18h and 78 h. In contrast, the comparison indicated that the mean differences were not significant ($p > 0.05$) at the late stage of larvae development. Hence, it was believed that diclofenac only gave an impact on larvae length between 18 h and 78 h of exposure. From the study, forensic entomologist might give a wrong assumption of PMI if they did not encounter the presence of drug when dealing with intoxication cases involving diclofenac fatality especially in the first 78 h of death.

In contrast, the presence of diclofenac did not have high impact on larvae width compared with the control as shown in Figure 11.2. There was no difference in the mean of width between control and treated groups at the early and late stage of exposure ($p > 0.05$). The width of larvae possessed significant difference ($p < 0.05$) only between control and treatment groups in the period of 48 h - 84 h of exposure. Statistical analysis within the treatment group larvae revealed irregular and fluctuation which only insignificantly observed ($p > 0.05$) between 60 h and 72 h exposure, especially in the highest dose of diclofenac.

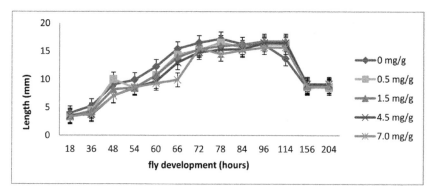

FIGURE 11.1   Mean of length of *Chrysomya megacephala* larvae under the influence of diclofenac diet.

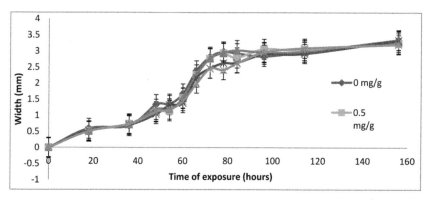

FIGURE 11.2   Mean of width of *Chrysomya megacephala* larvae under the influence of diclofenac diet.

11.3.2   EFFECT OF DICLOFENAC ON TOTAL DEVELOPMENT OF CHRYSOMYA MEGACEPHALA

In this study, we observed that there was a slight deviation on total development of *C. megacephala* in the presence of diclofenac. Data obtained from Table 11.1 showed that the development of first instar larvae was similar between control and treated larvae. However, the development of larvae containing diclofenac began to delay at the second instar larvae stage onwards for all doses especially in the highest dose (7.0 mg/g). From the study, it was found that the total development of the control completed within 198 h, earlier compared with the highest dose of diclofenac (7.0 mg/g) which was delayed to 222 h. Generally, the lower doses of diclofenac also affected the larvae development which slowed down the total duration to 6 h. Meanwhile, the presence of highest dose of diclofenac delayed the *C. megacephala* development up to 24 h, thus it may have an impact on the estimation of PMI.

TABLE 11.1   Duration (hours) of Blowfly *Chrysomya megacephala* Development under the Influence of Different Doses of Diclofenac

Stage of Larvae Development	Doses of Diclofenac				
	0 mg/g	0.5 mg/g	1.5 mg/g	4.5 mg/g	7.0 mg/g
First instar larvae	6- 30	6- 36	6- 36	6- 36	6- 36
Second instar larvae	36- 48	42- 54	42- 54	42- 60	42- 66
Third instar larvae	54- 102	60- 120	60- 120	66- 126	72- 132
Pre-pupa	108- 114	126- 132	126- 132	132- 138	138- 144
Pupae	120- 192	138- 198	138- 204	144- 210	150- 216
Adult	198	204	210	216	222

Data represent the total duration of *C. megacephala* to develop from first instar larvae until emergence of adult reared on liver spiked with different doses of diclofenac.

Several attempts have been made using *C. megacephala* to examine the effect of drug and toxins in its development. The study by Liu et al. [29] reported that duration of larvae and pupa development in the presence of malathion for both muscle and liver feeding was slower than control group. Thus, altering the PMI estimation by up to 36 h in muscle and deviated by up to 28 h in liver spiked with malathion. Similarly, Rashid et al. [26] also revealed that the development rate of *C. megacephala* was altered in the presence of malathion which was slower compared with control.

In contrast, Rashid et al. [30] also reported that ketum extract had an impact on the development rate of *C. megacephala* which attained the maximum length earlier than control after 36 h of exposure. However, the development rate of *C. megacephala* was almost similar to control when exposed to gun-shot residue (GSR) and the length of test larvae only affected to attain minimum size after longer exposure to GSR [31].

11.3.3 QUALITATIVE AND QUANTITATIVE ANALYSIS OF DICLOFENAC FROM BLOWFLY SAMPLE

Qualitative analysis confirmed the presence of diclofenac in blowfly samples at all treatment doses. The quality control was done by comparing the standard chromatogram of diclofenac with chromatogram of the sample. The presence of diclofenac was observed at retention time of 2.972 ($\pm$0.01) min in both standard and sample chromatograms as shown in Figure 11.3. In addition, the limit of detection (LOD) and limit of quantification (LOQ) were determined from the series of working solution of standard which was 0.042 mg/L and 0.25 mg/L, respectively.

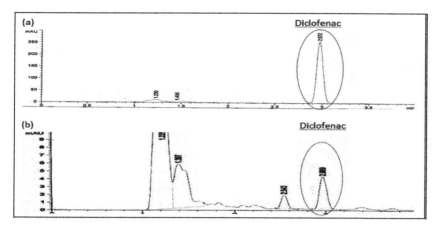

FIGURE 11.3 Representative chromatogram for (a) standard of diclofenac and (b) blowfly sample of third instar larvae collected from liver spiked with diclofenac sodium at 7.0 mg/g.

Based on the calibration curve in Figure 11.4, the amount of diclofenac contained in the flies was calculated. Table 11.2 shows the concentration of diclofenac in fly samples. The highest concentration of diclofenac was observed in samples collected from the highest dose of diclofenac-spiked liver. Overall, the concentrations of diclofenac detected in larvae sample at all doses ranged from 5.90 μg/g to 1468.00

μg/g. There was similar trend on fluctuation of diclofenac concentrations present in fly sample which was highest in third instar and prepupae larvae. This is probably due to accumulation of diclofenac in *C. megacephala* throughout its development [32]. After pre-pupae, the concentration of diclofenac seemed to decline gradually. This is because the drug might have passed through the metabolism process. However, the pharmacokinetics studies in blowfly larvae system have yet to be well understood.

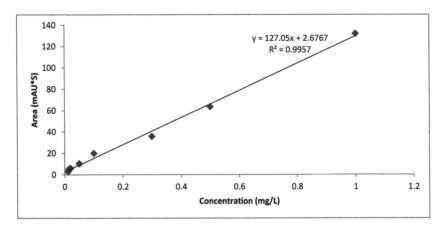

FIGURE 11.4   Standard calibration curve of diclofenac.

TABLE 11.2   Concentration of Diclofenac (μg/g) detected in *Chrysomya megacephala* Larvae Sample at Different Doses.

Stage of Blowfly Development	Concentration of Diclofenac (μg/g)			
	0.5 mg/g	1.5 mg/g	4.5 mg/g	7.0 mg/g
First instar larvae	11.70 ± 0.123	9.10 ± 0.008	7.30 ± 0.012	10.28 ± 0.055
Second instar larvae	5.90 ± 0.470	19.60 ± 0.218	58.36 ± 0.087	67.10 ± 0.120
Third instar larvae	132.40 ± 0.898	581.50 ± 0.387	1310.50 ± 0.118	1468.00 ± 0.086
Pre-pupa	47.50 ± 11.308	69.10 ± 0.144	474.80 ± 0.102	252.60 ± 0.074
Pupae	17.10 ± 0.379	8.40 ± 0.136	73.50 ± 0.075	72.20 ± 0.087
Adult	13.70 ± 0.175	24.10 ± 0.163	90.20 ± 0.066	15.43 ± 0.033

Data represent concentration of diclofenac detected in *C. megacephala* samples ($n = 5$) collected from liver spiked with different doses of diclofenac.

This positive result would prove that the entomological sample especially blowfly could be used in determination of the presence of toxicant from mortal intoxication cases to replace the absence of biological samples at the crime scene. The detection of diclofenac in blowfly *C. megacephala* sample explained the theory of ingestion and accumulation of drug in larvae throughout their development [5–33].

11.4 CONCLUSION

From this study, we can conclude that the presence of diclofenac has altered the development rates of blowfly *C. megacephala* up to 24 h at highest dose, thus it has an impact on estimation of PMI.

The study also revealed the potential use of blowfly larvae as alternative sample collected at crime scene for toxicological analysis to replace human tissue especially when biological samples were absent at the crime scene. These findings would provide valuable clue to forensic investigator, in solving such cases involving diclofenac intoxication especially on estimation of the PMI.

ACKNOWLEDGMENT

The authors are grateful to Institute of Forensic Science and Faculty of Applied Sciences, Universiti Teknologi MARA for providing research facilities and financial support.

KEYWORDS

- diclofenac
- *Chrysomya megacephala*
- postmortem interval
- entomotoxicology

REFERENCES

1. Byrd, J. H.; Castner J. L. *Forensic entomology: The Utility of Arthropods in Legal Investigations*; CRC Press: Boca Raton, Florida, 2001.
2. Amendt, J.; Krettek, R.; Zehner, R. Forensic Entomology. *Naturwissenschaften* **2004**, *91*, 51–65.
3. Greenberg, B.; Kunich, J. C. *Entomology and the Law—Flies as Forensic Indicators*. Cambridge University Press: Cambridge, U.K., 2002.
4. Goff, M. L.; Lord, W. D. Entomotoxicology, Insects as Toxicological Indicators and the Impact of Drugs and Toxins on Insect Development. In *Forensic Entomology: The Utility of Arthropods in Legal Investigation*; Byrd, J. H.; Castner, J. L., 2nd ed. CRC Press: Boca Raton, FL, 2010, 427–436.
5. Goff, M. L. Early Post-Mortem Changes and Stages of Decomposition in Exposed Cadavers. *Exp. App. Acarol.* **2009**, *49*, 21–36.
6. Anderson, G. S. Minimum and Maximum Development Rates of Some Forensically Important Calliphoridae (Diptera). *J. Forensic Sci.* **2000**, *45*(4), 824–832.
7. Sharma, R.; Garg, K. R.; Gaur, J. R. Contribution Of Various Measures for Estimation of Post Mortem Interval from Calliphoridae: A Review. *Egypt J. Forensic Sci.* **2013**, *5*(1), 1–12.
8. Introna, F.; Campobasso, C. P.; Goff, M. L. Entomotoxicology. *Forensic Sci Int.* **2001**, *120*, 42–47.
9. Campobasso, C. P.; Gherardi, M.; Caligara, M.; Sironi, L.; Introna. F. Drug Analysis in Blowfly Larvae and in Human Tissues: A Comparative Study. *Int J. of Leg Med.* **2004**, *118*, 210–214.
10. Monthei, D.R. Entomotoxicological and Thermal Factors Affecting the Development of Forensically Important Flies. Virginia Polytechnic Institute and State University: Blacksburg, 2009.
11. Hamid, N. A.; Omar, B.; Marwi, M. A.; Salleh, A. F. M.; Mansar, A. H.; Feng, S. S; Mokhtar, N. A. Review of Forensic Specimens Sent to Forensic Entomology Laboratory, Universiti Kebangsaan Malaysia for the year 2001. *Trop. Biomed.* **2003**, *20*, 27–31.
12. Sukontason, K.; Narongchai, P.; Kanchai, C.; Vichairat, K.; Sribanditmongkol, P.; Bhoopat, T. Forensic Entomology Cases in Thailand: A Review of Cases from 2000 to 2006. *Parasitol Res.* **2007**, *101*(5), 1417–1423.
13. Ahmad, N. W.; Lim, L. H.; Dhang, C. C.; Chin, H. C.; Abdullah, A. G.; Mustaffa, W. N. W.; Kian, C. W.; Jeffery, J.; Hashim, R; Azirun, S. M. Comparative Insect Fauna Succession on Indoor and Outdoor Monkey Carrions in a Semi-forested Area in Malaysia. *Asian Pac. J. Trop. Biomed.* *1*(2), **2011**, S232–S238.
14. Kumara, T; Ahmad, A. H.; Rawi, C. S. M.; Singh, B. Growth of *Chrysomya megacephala* (Fabricius) Maggots in a Morgue Cooler. *J. Forensic Sci.* **2010**, *55*(6), 1656–1658.
15. Ahmad, A.; Ahmad, A. H. A Preliminary Study on the Decomposition and Dipteran Associated with Exposed Carcasses in an Oil Palm Plantation in Bandar Baharu, Kedah, Malaysia. *Trop. Biomed.* **2009**, *26*(1), 1–10.
16. 16. Heo, C. C.; Mohamad, A. M.; Ahmad, F. M. S.; Jeffrey, J.; Omar, B. A. Preliminary Study of Insect Succession on a Pig Carcass in a Palm Oil Plantation in Malaysia. *Trop. Biomed.* **2007**, *24*(2), 23–27.

17. Boelsterli, U. A. Diclofenac-induced Liver Injury: A Paradigm of Idiosyncratic Drug Toxicity. *Toxicol. App. Pharmacol.* **2003**, *192*, 307–322.
18. Banks, A.; Zimmerman, H. J.; Ishak, K. G.; Harter J. G. Diclofenac-Associated Hepatotoxicity: Analysis of 180 Cases Reported to the Food and Drug Administration as Adverse Reactions. *Hepatology* **1995**, *22*, 820–882.
19. Hersh, E. V.; Pinto, A.; Moore, P. A. Adverse Drug Interactions Involving Common Prescription and Over-the-Counter Analgesic Agents. *Clin. Ther.* **2007**, *29*, 2477–2497.
20. Ng, L. E.; Halliwell, B.; Wong, K. P. Nephrotoxic Cell Death by Diclofenac and Meloxicam. *Biochem Biophys. Res. Commun.* **2008**, *369*, 873–877.
21. Han, E.; Kim, E.; Hong, H.; Jeong, S.; Kim, J. Evaluation of Post-Mortem Redistribution Phenomena for Commonly Encountered Drugs. *Forensic Sci. Int.* **2012**, *219*, 265–271.
22. Swan, G. E.; Cuthbert, R; Quevedo, M; Green, R. E.; Pain, D.J.; Bartels, P.; Cunningham, A. A.; Duncan, N.; Meharg, A.A.; Oaks, J. L.; Parry-Jones, J.; Shultz, S.; Taggart, M. A.; Verdoorn, G.; Wolter, K. Toxicity of Diclofenac to Gyps Vultures. *Bio. Lett.* **2006**, *2*, 279–282.
23. Ogada, D. L.; Keesing, F.; Virani, M. Z. Dropping Dead: Causes and Consequences of Vulture Population Declines Worldwide. *Ann. N.Y. Acad. Sci.* **2011**, *1249*, 1–15.
24. Green, R. E.; Newton, I.; Shultz, S.; Cunningham, A. A.; Gilbert, M.; Pain, D. J.; Prakash, V. Diclofenac Poisoning as a Cause of Vulture Population Declines Across the Indian Subcontinent. *J. App. Ecol.* **2004**, *41*, 793–800.
25. Beyer, J. C.; Enos, W. F.; Stajie, M. Drug Identification Through Analysis of Maggots. *J. Forensic Sci.* **1980**, *25*, 411–412.
26. Rashid, R. A; Khairul, O; Ismail, M. I; Zuha, R. M; Rogayah A. H. Determination of Malathion Levels and the Effect of Malathion on the Growth of *Chrysomya megacephala* (Fibricius) in Malathion-Exposed Rat Carcass. *Trop. Biomed.* **2008**, *25*(3), 184–190.
27. George, K. A.; Archer, M. S.; Green, L. M.; Conlan, X. A.; Toop, T. Effect of Morphine on the Growth Rate of *Calliphora stygia* (Fabricius) (Diptera: Calliphoridae) and Possible Implications for Forensic Entomology. *Forensic Sci. Int.* **2009**, *193*, 21–25.
28. Adam, Z. J. O.; Hall, M. J. R. Methods Used for the Killing and Preservation of Blowfly Larvae and Their Effect on Post-Mortem Larval Length. *Forensic Sci. Int.* **2003**, *13*, 50–61.
29. Liu, X.; Shi, Y.; Wang, H.; Zhang, R. Determination of Malathion Levels and Its Effect on the Development of *Chrysomya megacephala* (Fabricius) in South China. *J. Forensic Sci. Int.* **2009**, *192*, 1–3.
30. Rashid, R. A.; Zulkifli, N. F.; Rashid, R. A.; Rosli, S. F.; Sulaiman, S. H. S.; Nazni, W. A. Effects of Ketum Extract on Blowfly *Chrysomya megacephala* Development and Detection of Mitragynine in Larvae Sample. In IEEE Symposium on Business, Engineering and Industrial Applications, **2012**, 337–341.
31. Rashid, R. A.; Khairul, O.; Zuha, R. M.; Heo, C. C. An Observation on the Decomposition Process of Gasoline Ingested Monkey Carcasses in a Secondary Forest in Malaysia. *Trop. Biomed.* **2008**, *27*(3), 373–383.
32. Sadler, D. W.; Fukeb, C.; Court, F.; Pounder, D. J. Drug Accumulation and Elimination in *Caliphora vicina* Larvae. *Forensic Sci. Int.* **1995**, *71*, 191–197.
33. Tracqui, A.; Keyser-Tracqui, C.; Kintz, P.; Ludes, B. Entomotoxicology for the Forensic Toxicologist: Much Ado About Nothing? *Int. J. Leg. Med.* **2004**, *118*, 194–196.

CHAPTER 12

THE EFFECT OF MANDIBULAR ANGULATION ON GONIAL ANGLE AND TOOTH LENGTH MEASUREMENT OF DENTAL PANORAMIC RADIOGRAPHS

TONG WAH LIM[1,2], BUDI ASLINIE MD SABRI[3], and ROHANA AHMAD[1,2,*]

[1]Centre for Restorative Dentistry Studies, Faculty of Dentistry, Universiti Teknologi MARA, 47000 Sg Buloh, Selangor, Malaysia

[2]Pharmacogenomic Institute (iPromise), Puncak Alam, Selangor, Malaysia

[3]Centre for Population Oral Health and Clinical Prevention Studies, Faculty of Dentistry, Universiti Teknologi MARA, 47000 Sg Buloh, Selangor, Malaysia

*Corresponding author. E-mail: drrohana@salam.uitm.edu.my

CONTENTS

ABSTRACT

The aim of this study was to assess the validity of dental panoramic radiographs as a diagnostic and evaluation tool for dental implantology and removable prosthodontics. One fully dentate dry mandible was used as the radiographic model. Radiographs of the mandible were taken at various mandibular angulations. Two evaluation parameters, the gonial angle and tooth length, were determined using these radiographs. The results showed that the measurement of gonial angles remained constant at various mandibular angulations while the length of canines and second premolars significantly decreased with increasing mandibular angulations raised by 4 mm increment from 4 to 28 mm. In conclusion, panoramic radiograph technique is suitable for gonial angle determination but not for bone height measurement especially in the anterior region of the mandible where distortion was most pronounced.

12.1   INTRODUCTION

The use of dental panoramic radiographs for screening and diagnostic purposes has been widely accepted and routinely practiced in all disciplines of dentistry. These radiographs are commonly used in dental research as they are likely to be found in records going back several years and therefore they constitute a ready source of data for retrospective study. However, the usefulness of panoramic radiographs for implant planning and research purposes is rather limited due to the difficulty in controlling the distortion and magnification errors associated with the images.

Farman et al. reported that panoramic images of the same patient made at intervals are impossible to superimpose due to distortion from variable head positions [1]. Some other studies revealed a 10% magnification error associated with these radiographic images [2, 3]. A few studies reported that the distortion could be reduced by applying a magnification factor to measurement, comparing proportions rather than actual measurements and using proportional area index [3–5]. Despite all these efforts to reduce the distortion and magnification errors of panoramic radiograph, this two-dimensional (2-D) imaging technique may not be a reliable measurement tool for research and diagnostic purposes in dentistry.

Batenburg et al. and Stellingsma reported that mandibular angulations as a result of variable intermaxillary distance in edentulous patient can affect the magnification factors of the produced radiograph significantly [6, 7]. These edentulous patients may not have dentures to bite on or the denture teeth may have worn considerably resulting in reduced intermaxillary distance or mandibular angulation. Unless meticulous precautions are taken to reproducibly position the edentulous patient at the exact mandibular angulation in the X-ray apparatus during radiographic acquisition in subsequent years, a 2-D radiograph may be flawed with accuracy issues and may not be a reliable quantitative measurement technique for dental research and clinical practice. To date, several techniques of measurement of three-dimensional (3-D) images from cone beam computed tomography (CBCT) based on a series of anatomical landmarks have been commonly used and reported to provide more accurate and reproducible results [8, 9]. However, due to the cost and technical issues of CBCT, most general dental practitioners in Malaysia are still relying on panoramic radiographs as an evaluation tool in implant dentistry without taking into consideration the effect of inconsistent mandibular angulation of the edentulous patients.

The aim of this study was to assess the validity of dental panoramic radiographs as a diagnostic and evaluation tool in implantology and removable prosthodontics. In particular, the effect of various mandibular angulations on tooth length and gonial angle on the panoramic radiographs was evaluated.

12.2   MATERIALS AND METHODS

This study was carried out in the Radiology Unit, Faculty of Dentistry, Universiti Teknologi MARA. A fully dentate dry mandible was used in this study as the radiographic model. A dental panoramic radiograph was taken with this mandible positioned in the orthopantomographic machine similar to a patient's head position. Both the mandibular angles were supported using predetermined thickness of beading waxes and the symphysis region was secured by the chin brace anteriorly. The symphysis midline of the mandible was positioned at the center using the guide of light cross of the machine.

The initial mandibular angulation was set by raising the angle of the mandible on both sides by 4 mm by using a predetermined thickness of beading wax. Raising (tilting) the mandible at the angles simulates mouth opening of the patient; the greater the tilt, the bigger the mouth opening. The angulation was subsequently increased by 4 mm by adding more beading wax. Consequently, the angles were raised posteriorly at

8, 12, 16, 20, 24, and 28 mm. A dental panoramic radiograph was taken for each of the mandibular angulations (Planmeca ProMax 3D, Planmeca Oy, Helsinki) (Fig. 12.1). An exposure of 74 kV, 9mA, and 16 s was standardized for all radiographs. The images were processed using Kodak DryView 6800 Laser Imager.

A tracing paper was superimposed on the panoramic radiograph placed on an X-ray viewer and the radiograph was traced based on a series of anatomical landmarks using sharp pencil. The gonial angle (in degrees) was determined using a mathematical protractor at the angle of the mandible. It was determined from the intersection of two tangents whereby the first line was drawn by connecting the two most posterior points on the posterior border of the ramus and the second line was drawn by connecting the two most inferior points on the lower border of the body of mandible.

The crest of the alveolar bone could not be clearly delineated on the radiographs due to overlapping of tooth structures and therefore to simulate the measurement of bone height in two locations, anterior and posterior, the length of two teeth, the canine and premolar were calculated, respectively. The tooth length (in mm) was measured from the tip of the crown to the tip of the root using a millimeter metal ruler. The mental foramina were also difficult to delineate and hence were not used as landmarks for measurement of bone height.

12.3   RESULTS AND DISCUSSION

The gonial angle values remain relatively constant despite the change in the mandibular angulation from 4 mm to 28 mm. The difference did not exceed 2° (Table 12.1) at the various mandibular angulations. In contrast, the tooth length of both canines and second premolars demonstrated significant decrease in length with each increase in the mandibular angulations (Tables 12.2 and 12.3). The biggest difference in canine and second premolar tooth length was 10.5 and 3 mm, respectively, when the mandibular angulation was raised from 4 mm to 28 mm.

The relationship between tooth length and mandibular angulation was further investigated using Pearson product-moment correlation coefficient. Preliminary analyses were performed to ensure no violation of the assumptions of normality, linearity, and homoscedasticity. There was a strong, negative correlation between right canine tooth length ($r = -0.99$, $n = 7$, $p < 0.001$), left canine tooth length ($r = -0.96$, $n = 7$, $p < 0.001$), right

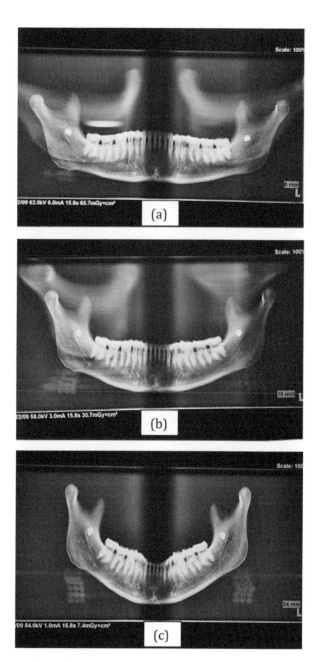

FIGURE 12.1 Three panoramic radiographs taken at various mandibular angulations: (a) 8 mm, (b) 16 mm, and (c) 24 mm.

premolar tooth length ($r = -0.94$, $n = 7$, $p = 0.001$), and left premolar tooth length ($r = -0.93$, $n = 7$, $p = 0.003$) with mandibular angulation where shorter tooth length is associated with larger mandibular angulation.

TABLE 12.1   Gonial Angles at Various Mandibular Angulations.

Mandibular Angulation (mm)	Right Gonial Angle (Degree)	Left Gonial Angle (Degree)
4	124	126
8	123	126
12	124	126
16	125	126
20	124	126
24	124	125
28	125	124

TABLE 12.2   Mandibular Canine Tooth Length at Various Mandibular Angulations.

Mandibular Angulation (mm)	Right Canine Tooth Length (mm)	Left Canine Tooth Length (mm)
4	25.0	24.5
8	25.0	23.5
12	23.0	21.5
16	19.5	19.0
20	19.0	18.0
24	17.0	16.0
28	17.0	14.0

TABLE 12.3   Mandibular Second Premolar Tooth Length at Various Mandibular Angulations.

Mandibular Angulation (mm)	Right 2nd Premolar Tooth Length (mm)	Left 2nd Premolar Tooth Length (mm)
4	24.0	22.5
8	24.0	22.5
12	24.0	22.0
16	23.5	22.0
20	23.0	21.5
24	21.5	21.0
28	21.0	20.0

The measurement of gonial angle is important as this angle plays a significant role in influencing the amount of bone resorption that would occur underneath a denture [10]. Patients with low gonial angle have been shown to suffer more bone resorption compared with patients with high gonial angle. This angle was commonly measured using lateral cephalometric radiographs. However, Oksayan et al. reported that there were no significant differences in the magnitude of the gonial angles determined by either panoramic or lateral cephalometric radiographs [11]. As shown in this study, the gonial angle determination using panoramic radiograph is not influenced by the mandibular angulations. Hence, the use of panoramic radiograph for measurement of gonial angle and its changes over time is recommended.

The number of elderly population is growing rapidly in Malaysia and most of them require dentures to chew effectively. Many individuals cannot tolerate the looseness of the mandibular dentures; dental implants are increasingly used in the canine and/or premolar regions to help retain the dentures. Hence is the rationale for choosing mandibular canines and second premolars for tooth length measurement to simulate bone height in this region.

This study has shown that changing the mandibular angulations significantly influences the radiographic tooth length measurement for canines and second premolars. As the mandibular angulation enlarges (increases mouth opening), the tooth length shortens. The effect is more pronounced with the canine as it is located more anteriorly in the mandible. This result is consistent with the results reported by Batenburg et al. which showed that the anterior part of the mandible suffers greater distortion compared with the posterior region of the mandible [6]. The shortening error can be up to 42% in magnitude as demonstrated by measurement of the left canine. It can be inferred that bone height measurement in this region will be similarly affected by the mandibular angulation. Therefore, panoramic radiograph should be used cautiously to measure bone height in edentulous patient as the bone may appear higher or shorter depending on the angulation of the mandible. Error in measurement of the bone height may result in serious complications including damage to neurovascular structure during implant surgery and biomechanical failures of the prostheses. Therefore, it is recommended that other radiographs such as periapicals or 3-D images should be used to measure bone height before implant placement.

12.4 CONCLUSION

In summary, the mandibular angulation during the acquisition of panoramic radiograph has no effect on the gonial angle but it affects the tooth length measurement significantly and is more pronounced for the anterior teeth.

ACKNOWLEDGMENT

This study was funded by Universiti Teknologi MARA RIF grant 600-RMI/DANA 5/3/RIF (570/2012) and Ministry of Education Malaysia RAGS grant 600-RMI/RAGS 5/3 (117/2013).

KEYWORDS

- gonial angle
- mandibular angulation
- tooth length
- radiograph
- implant dentistry

REFERENCES

1. Farman, A. G.; Phelps, R.; Downs, J. B. Artifact or Pathosis? Problem-solving Panoramic Dental Radiology (I). *Quintessence Int. Dent. Dig.* **1983,** *14* (1), 55–65.
2. Christen, A. G.; Segretti, V. A. Distortions and Artifacts Encountered in Panorex. *J. Am. Dent. Assoc.* **1968,** *77,* 1096–1101.
3. Wilding, R. J. C.; Levin, I.; Pepper, R. The Use of Panoramic Radiographs to Measure Alveolar Bone Areas. *J. Oral Rehabil.* **1987,** *14,* 557–567.
4. van Waas, M. A. J. Ridge Resorption in Denture Wearers after Vestibuloplasty and Lowering of the Floor of the Mouth, Measured on Panoramic Radiographs. *Dentomaxillofac. Radiol.* **1983,** *12,* 115–121.
5. Wical, K. E.; Swoope, C. C. Studies of Residual Ridge Resorption. Part I. Use of Panoramic Radiographs for Evaluation and Classification of Mandibular Resorption. *J. Prosthet. Dent.* **1974,** *32,* 7–12.
6. Batenburg, R. H. K.; Stellingsma, K.; Raghoebar, G. M.; Vissink, A. Bone Height Measurements on Panoramic Radiographs. The Effect of Shape and Position of

Edentulous Mandibles. *Oral Surg. Oral Med. Oral Pathol. Oral Radiol. Endod.* **1997,** *84,* 430–435.

7. Stellingsma, K.; Batenburg, R. H. K.; Meijer, H. J. A.; Raghoebar, G. M.; Kropmans, T. J. B. The Oblique Radiographic Technique for Bone Height Measurements on Edentulous Mandibles. A Preclinical Study and an Introduction to the Clinical Use. *Oral Surg. Oral Med. Oral Pathol. Oral Radiol. Endod.* **2000,** *89,* 522–529.

8. Tai, K.; Park, J. H.; Mashima, K.; Hotokezaka, H. Using Superimposition of 3-Dimensional Cone-beam Computed Tomography Images with Surface-based Registration in Growing Patients. *J. Clin. Pediatr. Dent.* **2010,** *34,* 361–367.

9. Ahmad, R.; Abu-Hassan, M. I.; Li, Q.; Swain, M. V. Three Dimensional Quantification of Mandibular Bone Remodeling Using Standard Tessellation Language Registration Based Superimposition. *Clin. Oral Implants Res.* **2013,** *24,* 1273–1279.

10. Tallgren, A. The Continuing Reduction of the Residual Alveolar Ridges in Complete Denture Wearers: A Mixed-longitudinal Study Covering 25 Years. *J. Prosthet. Dent.* **1972,** *27,* 120–132.

11. Oksayan, R.; Aktan, A. M.; Sökücü, O.; Hastar, E.; Ciftci, M. E. Does the Panoramic Radiography Have the Power to Identify the Gonial Angle in Orthodontics? *Sci. World J.* **2012,** 1–4.

PART III
Renewable Products from Agricultural and Natural Sources

CHAPTER 13

SCREENING OF THE BEST CARBON AND NITROGEN SOURCES FOR *PSEUDOMONAS AERUGINOSA* AS A POTENTIAL POLYHYDROXYALKANOATES (PHAs) AND RHAMNOLIPIDS PRODUCER

MAZNI MOHD YATIM[1],
TENGKU ELIDA TENGKU ZAINAL MULOK[1,*],
NIK ROSLAN NIK ABDUL RASHID[1], and AMIZON AZIZAN[2]

[1]*Department of Biology, Faculty of Applied Science, Universiti Teknologi MARA, Shah Alam, Malaysia*

[2]*Department of Chemical Engineering, Faculty of Chemical Engineering, Universiti Teknologi MARA, Shah Alam, Malaysia*

Corresponding author. E-mail: tetzm@salam.uitm.edu.my

CONTENTS

ABSTRACT

Soil samples were collected from oil-polluted area and three isolates were screened as potential producer of PHAs and rhamnolipids. The identity of the isolates was confirmed as *P. aeruginosa* using biochemical tests and 16S rRNA sequencing. The main objective of this study was to screen for the best combination of carbon and nitrogen sources that support the optimal growth of this strain using predetermined medium at pH 6.8. The medium composition was 3.0 g KH_2PO_4, 7.0 g K_2HPO_4, and 0.2 g $MgSO_4.7H_2O$ with addition of different combinations of carbon and nitrogen sources for different runs of fermentation using shake flask scale, at temperature 37°C, as suggested by the full factorial design. From the ANOVA, the significant factors that contributed to the optimum growth of *P. aeruginosa* were palm oil as carbon source and ammonium nitrate as nitrogen source. The highest value of cell dry weight (CDW) was achieved at 5.22 g/L comprising of 5% (w/v) of each carbon source and 3% (w/v) of each nitrogen source. Further study on optimization of both PHAs and rhamnolipids using response surface methodology (RSM) is suggested.

13.1 INTRODUCTION

Polyhydroxyalkanoates (PHAs) are the compounds that consist of both carbon and energy that can be found intracellularly. They are produced by many bacterial strains in the form of inclusion bodies [1]. PHAs are also known as microbial storage polyester that can be synthesized naturally using various carbon substrates by many strains of bacteria via microbial fermentation [2–4]. They also stated that PHAs synthesized by *Pseudomonas aeruginosa* as carbon and energy storage compounds are co-polyesters principally composed of medium-chain-length (MCL) monomers with carbon chain length ranging from 6 to 14 carbon atoms. This monomer composition depends on the PHA syntheses, the types of the carbon source, and the metabolic pathways involved. PHAs have attracted commercial biotechnological interest due to their biodegradability and biocompatibility [5].

Rhamnolipids are amphiphilic molecules that are typically constituted of MCL (R)-3-hydroxyfatty acids known as 3-hydroxyalkanoate, linked through a β-glycosidic bond to mono- or di-rhamnose moiety [6]. It is also a monomer unit of MCL PHAs. Therefore, (MCL) PHAs and rhamnolipids have 3-hydroxyalkanoates as common structural units; the most typical one is 3-hydroxydecanoate [7].

PHAs are synthesized intracellularly; hence, the recovery of PHAs requires the disruption of the bacterial cells that contribute to high production cost and a large portion of the total cost of fermentation process [8]. One potential approach for reducing the production cost of both products is by producing both metabolites simultaneously as was reported by Hori et al. [8] and Marsudi et al. [9]. The cells can be used for the production of rhamnolipids since they are secreted extracellularly before eventual disruption to recover PHAs.

P. aeruginosa was chosen in this study since not many bacterial species are able to produce both rhamnolipids and PHAs. Majority of the bacterial species can either produce only rhamnolipids or only PHA but not both.

In this study, fermentation process was done in a shake flask scale since it is easy to handle and widely applied in academic field and bioindustry for screening and bioprocess development projects [10–13]. Besides that, since the duration of fermentation process was only 3 days for *P. aeruginosa*, there is no worry about the oxygen limitation for this culture.

The aim of this study was to isolate *P. aeruginosa* from oil-polluted area as it is a potential producer of PHAs and rhamnolipids and to screen the carbon and nitrogen sources that support the optimum growth of this bacterial strain using predetermined medium. The screening step is important before conducting the optimization process since the concentration of the carbon and nitrogen sources to be evaluated can be focused.

13.2 MATERIALS AND METHODS

13.2.1 ISOLATION AND IDENTIFICATION OF BACTERIAL ISOLATES

An indigenous microorganism has been selected among many bacterial species isolated from oil-polluted zone collected from car workshop due to its remarkable ability for reducing water surface tension. The soil samples were collected from three car workshops in Shah Alam, Selangor, Malaysia. The soil was later diluted to enumerate the microorganisms. For preliminary identification, the isolate was cultured on Pseudo F agar and identified as *P. aeruginosa*. In order to further confirm the identity of the isolates as *P. aeruginosa*, biochemical tests were carried out according to

Bergey's manual. After successful screening, the isolate identified as *P. aeruginosa* was inoculated into 70% (v/v) glycerol as stock culture and for maintaining the cells condition. The selected isolate then underwent further molecular work for 16S rRNA sequencing in order to confirm its identity.

13.2.2 DETERMINATION OF BIOSURFACTANT PRODUCTION

The presence of biosurfactant in the medium after several days of fermentation was determined using several methods namely drop collapse, pendant drop, and emulsification tests.

The drop collapse test was carried out according to Bodour et al. [14]. The culture broth shaking at 180 rpm for 7 days was initially supplemented with carbon source but on the third day of fermentation, 0.1% (v/v) of crude oil was added. After 7 days, the broth was harvested, centrifuged at 15,000 rpm for 25 min. The supernatant at a volume of 0.1 mL was later dispensed into the wells of a sterilized white tile, priorly added with 1 mL of sterilized liquid paraffin. After 10 min, the well was observed visually for the break in the surface tension of the paraffin.

The oil displacement test was performed by referring to Anandaraj and Thivakaran [15] with some modifications. Distilled water of 30 mL was poured onto sterile petri dishes and allowed to spread evenly, followed by addition of 1 mL of crude oil to the center. Supernatant of 20 μl volume, obtained from bacterial cultures after centrifugation at 10,000 rpm for 5 min was later added to the center of oil and water mixture. The biosurfactant thus produced can displace the oil and spread in the water. The diameter of the developed clear zone was measured. Distilled water of 1 μl volume added to the center of oil–water mixture was used as a negative control. The measurement of the clear zone was done in triplicates.

Emulsification index of the culture samples was determined by adding 2 mL of a hydrocarbon, which was kerosene, to the same amount of culture supernatant derived from centrifugation at 13,000 rpm. It was later mixed well using a vortex for 2 min and leaving to stand for 24 h. The E24 index was determined as percentage of the height of emulsified layer (mm) divided by total height of the liquid column (mm) as noted by Cooper and Goldenberg [16].

13.2.3 FERMENTATION USING SHAKE FLASKS FOR PHAS AND RHAMNOLIPIDS PRODUCTION

The medium composition was 3.0 g KH_2PO_4, 7.0 g K_2HPO_4, and 0.2 g $MgSO_4 \cdot 7H_2O$ with addition of different combinations of carbon and nitrogen sources for different runs of fermentation suggested by the full factorial design. The pH of the media was 6.8 and the temperature was 37°C. After 3 days of fermentation at 37°C, shaking at 150 rpm, the medium was harvested and centrifuged at 8000 rpm for 15 min for separation of pellet and supernatant. The pellet was used for PHAs determination, while the rhamnolipids content was determined from the supernatant.

13.2.4 EXTRACTION OF PHAS

The cell pellets were washed with distilled water and later dried at 90°C in the oven for 3 h. The cells were dried until a constant weight was achieved. A suspension (9 mL methanol + 0.3 mL H_2SO_4) of 0.6 mL volume was later added into the vials containing 0.01 g of dried cell pellets and was dried in the oven for another 2 h. The cell pellets were then cooled at room temperature before the addition of 0.6 mL of chloroform and 0.3 mL of distilled water. The mixtures were mixed slowly to allow cell separation. Centrifugation was carried out at 8000 rpm for 1 min, after which the top chloroform layer was collected and transferred into a sterile vial. The sample was later analyzed for the presence of PHAs using Fourier transform infrared spectroscopy (FTIR).

13.2.5 EXTRACTION OF RHAMNOLIPIDS

The pH of the supernatant was adjusted to 2.0 and allowed to stand overnight at 4°C. A volume of 3 mL of supernatant was then extracted with a mixture of $CHCl_3:CH_3OH$ (2:1; v/v). The solvent was later evaporated and the residue formed was dissolved in 3 mL of 0.1 mol/L $NaHCO_3$.

The presence of rhamnolipids and PHAs were analyzed using quantitative and qualitative methods.

13.2.6 QUALITATIVE METHOD

The qualitative test for detecting the presence of PHA was performed by using Nile blue staining. PHA granules exhibited a strong fluorescence when stained with Nile blue A. Heat-fixed cells were treated with 1% (w/v) Nile blue A for 10 min and were observed under fluorescence microscope at an excitation wavelength of 460 nm.

Rhamnose test was used for rhamnolipids. This test was able to detect the presence of carbohydrate groups in the biosurfactant molecule. A mixture of 0.5 mL of cell supernatant, 0.5 mL of 5% (v/v) phenol solution, and 2.5 mL of sulfuric acid was incubated for 15 min. Absorbance was measured at 490 nm using the spectrophotometer.

The presence of rhamnolipids can also be detected by using thin-layer chromatography by employing anthrone reagent. The rhamnolipids extracted before were analyzed on silica gel plate. The silica gel plate was developed comprising a mixture of 65% (v/v) chloroform, 15% (v/v) methanol, and 2% (w/v) acetic acid. After the mobile phase reached the marked upper line of the plate, the plate was later sprayed with anthrone reagent. Yellow spot appearing on the plate is an indication of the presence of rhamnolipids. Sodium dodecyl sulfate (SDS) acted as positive control, while the negative control was sterile distilled water.

13.2.7 QUANTITATIVE METHOD

The presence of both PHA and rhamnolipids was further confirmed using FTIR. The KBr disc technique was employed as noted by Liu et al. (2011) [17]. Equal weight of both the sample and the KBr was used to compare the different functional chemical compounds present in the samples quantitatively.

13.3 RESULTS AND DISCUSSION

13.3.1 ISOLATION AND IDENTIFICATION OF BACTERIAL ISOLATES

Results from Gram staining showed that out of 21 isolates, only 10 were Gram negative. Only five of these Gram-negative isolates were oxidase negative, suggesting these isolates belonged to *Enterobacteriaceae*. The other five Gram-negative isolates which were oxidase positive (presumably *Aeromonas*, *Pseudomonas*, or *Vibrio* species) were further tested for glucose fermentation. Two out of these five isolates showed positive reaction, suggesting they belong to *Aeromonas* or *Vibrio* species, while the remaining three isolates showed negative reaction, suggesting *Pseudomonas* spp. These isolates were further grown on a selective media, Pseudo F agar, as a confirmatory test that is specific for Fluorescein producer. Thus, positive reaction obtained concluded that these isolates indeed belong to *Pseudomonas* spp.

The bacterial isolates were later grown on Pseudo P agar with added nonfluorescent diffusible blue pigment. This pigment selected for pyocyanin producer as suggested by Bergey's manual, serving as a conclusive test for confirmation of the identity of *P. aeruginosa*. All three isolates did show positive results, thus further confirming the identity of the three isolates as *P. aeruginosa*.

13.3.2 SCREENING FOR BIOSURFACTANT PRODUCTION

These isolates were later cultivated for 3 days to screen the presence of biosurfactant. All isolates showed positive results for drop collapse test compared with SDS acting as a positive control (Table 13.1). In the oil displacement test (Table 13.2), the biosurfactant produced by the isolates formed larger clear zone compared with the 1% (w/v) SDS acting as positive control. This indicated that the biosurfactant produced by these isolates have greater potential to compete with the synthetic surfactant, SDS.

TABLE 13.1 Results from Drop Collapse Test for Isolates A, B, and C.

Samples	Results
1 % (w/v) SDS	+
Uninoculated medium	–
Isolate A	+
Isolate B	+
Isolate C	+

Positive control: 1% SDS; negative control: uninoculated medium, +: presence of biosurfactant, –: absence of biosurfactant.

TABLE 13.2   Results from Oil Displacement Test for Isolates A, B, and C.

Samples	Diameter of Clear Zones (mm)
1 % (w/v) SDS	90
Uninoculated medium	–
Isolate A	90
Isolate B	90
Isolate C	90

Positive control: 1% SDS; negative control: uninoculated medium; –: absence of biosurfactant.

Emulsification index (E24) described the emulsification activity of the surfactant which was determined using Equation 13.1. The E24 for 1% (w/v) SDS, 68.75% (w/v); isolate A, 65.63% (w/v); isolate B, 68.75 % (w/v); and isolate C, 62.50% (w/v) (Table 13.3). The higher the E24 value, the better would be the emulsification activity of the surfactant.

$$E24 = \text{Measured emulsified layer height} \times 100\% \text{ total height.} \quad (13.1)$$

TABLE 13.3   Results from Emulsification Index for Isolates A, B, and C.

Samples	Percentage of Emulsified Layer (%)
1 % (w/v) SDS	68.75
Uninoculated medium	0.00
Isolate A	65.63
Isolate B	68.75
Isolate C	62.50

Positive control: 1% SDS; negative control: uninoculated medium.

The pendant drop measures the surface tension of the biosurfactant. The pendant drop is a drop suspended from a needle in a bulk liquid or gaseous phase. The shape of the drop results from the relationship between the surface tension or interfacial tension and gravity. The surface tension is calculated from the shadow image of a pendant drop using drop shape analysis.

The results, expressed in dynes/cm (Table 13.4), showed that the surface tension of 1% (w/v) SDS was higher than the surface tension demonstrated by the isolates. This indicated that the isolates have high potential as biosurfactant producer, including rhamnolipids. The surface activity of the bacteria can also be expressed as a percentage of the reduction in surface tension using Equation 13.2 [18, 19].

$$\text{Percentage of surface tension reduction} = \frac{(\tilde{a}m - \tilde{a}c) \times 100\%}{\tilde{a}m}, \quad (13.2)$$

where γm is the surface tension of the control (uninoculated medium) and γc is the surface tension of the tested supernatant of the isolates.

TABLE 13.4 Results from Pendant Drop Method for Isolates A, B, and C.

Samples	Surface Tension (dynes/cm)	Percentage of Surface Tension Reduction (%)
1% SDS	28.80	32.17
Uninoculated medium	42.46	–
Isolate A	21.13	50.24
Isolate B	19.76	53.46
Isolate C	23.41	44.87

Positive control: 1% SDS; negative control: uninoculated medium.

Among the three screening methods, the most accurate and precise result for comparing the strength of biosurfactant produced by these three isolates was by referring to the measurement of surface tension. From the table below, it is shown that isolate B produced the lowest surface tension value and the highest percentage of surface tension reduction. This means that this isolate was the best among the other isolates which was later sent for 16S rRNA sequencing and used in the fermentation study.

13.3.3 16S RRNA SEQUENCING

Isolate B then underwent 16S rRNA sequencing in order to confirm the status of the species. The search result from Basic Local Alignment Search Tool (BLAST) analysis revealed 99% identical to *P. aeruginosa* PAO1 strain. *P.aeruginosa* PAO1 is known to produce both PHAs and rhamnolipids. A metabolic link between PHAs and rhamnolipids syntheses has been suggested involving this strain [20, 6]. Choi et al. [4] and Hori et al. (2002) [8] reported the simultaneous production of both PHAs and rhamnolipids; thus, this strain was used as a comparative study in this finding.

13.3.4 FTIR ANALYSES TO SCREEN FOR THE PRESENCE OF PHAS AND RHAMNOLIPIDS

From the FTIR analyses, strong peaks appeared at 1728/cm, 1740/cm, 1732/cm and weak peaks also observed at 1280/cm, 1165/cm, and 2925/cm. This range of wave numbers coincided with the wave numbers belonging to PHAs (Table 13.5).

TABLE 13.5   FTIR Analyses Showing Significant Wave Number.

Sample	FTIR Wave Numbers
	1728/cm
	1740/cm
PHAs	1732/cm
	1280/cm
	1165/cm
	2925/cm

Several important peaks were also detected in rhamnolipid analyses representing C–OH group, O–H plane, O–C–O symmetric bond, C–O stretching, C–H deformation, CH_3 group, COO^- group, C=O bond of carboxylate group, symmetric O–H stretching, and O–H stretching of CH_2–CH_3 group (Table 13.6) as correlated with previous work of Christova et al. [21].

TABLE 13.6   FTIR Analyses Showing the Presence of Important Chemical Groups.

Samples	Chemical Groups Present
	C–OH group
	O–H plane
	O–C–O symmetric bond
	C–O stretching
Rhamnolipids	C–H deformation
	CH_3 group
	COO^- group
	C=O bond of carboxylate group
	Symmetric O–H stretching, O–H
	Stretching of CH_2–CH_3 group

13.3.5   SCREENING OF THE BEST CARBON AND NITROGEN SOURCES USING FULL FACTORIAL DESIGN

The shake flask fermentation was carried out for 21 runs using 5 center points as suggested by the full factorial design. The response used was the final dry weight of the fermentation products. The results for the 21 runs were tabulated in Table 13.7.

TABLE 13.7   Screening Results for Best Combination of Carbon and Nitrogen Sources using Full Factorial Design.

Run	Factors (%)						Cell Dry Weight (g/L)
	A	**B**	**C**	**D**	**E**	**F**	
1	7.00	7.00	7.00	8.00	8.00	8.00	0.71
2	7.00	7.00	7.00	8.00	8.00	8.00	0.75
3	7.00	7.00	7.00	8.00	8.00	8.00	0.76
4	7.00	7.00	7.00	8.00	8.00	8.00	0.72
5	7.00	7.00	7.00	8.00	8.00	8.00	0.73
6	5.00	5.00	5.00	3.00	3.00	3.00	5.22
7	9.00	5.00	5.00	3.00	13.00	3.00	4.14
8	5.00	9.00	5.00	3.00	13.00	13.00	1.96
9	9.00	9.00	5.00	3.00	3.00	13.00	5.02
10	5.00	5.00	9.00	3.00	13.00	13.00	0.99
11	9.00	5.00	9.00	3.00	3.00	13.00	2.85
12	5.00	9.00	9.00	3.00	3.00	3.00	2.70
13	9.00	9.00	9.00	3.00	13.00	3.00	4.32
14	5.00	5.00	5.00	13.00	3.00	13.00	2.83
15	9.00	5.00	5.00	13.00	13.00	13.00	0.39
16	5.00	9.00	5.00	13.00	13.00	3.00	1.49
17	9.00	9.00	5.00	13.00	3.00	3.00	2.19
18	5.00	5.00	9.00	13.00	13.00	3.00	0.21
19	9.00	5.00	9.00	13.00	3.00	3.00	1.09
20	5.00	9.00	9.00	13.00	3.00	13.00	3.03
21	9.00	9.00	9.00	13.00	13.00	13.00	0.13

The response ranged from 0.13 to 5.22 g/L. The highest value for the final dry weight was an indication of the best composition for the fermentation medium. According to the study by Choi et al. [4] and Hori et al.

(2011) [7] on *P. aeruginosa*, the cell dry weight (CDW) value is directly proportional to the PHAs and rhamnolipids contents. The higher the CDW, the higher would be the PHAs and rhamnolipids contents. It was assumed that during screening, the composition of the medium was the best to produce optimum PHAs and rhamnolipids when the highest CDW was achieved. The production of both PHAs and rhamnolipids will be focused during optimization in later study.

In this study, run 5 showed the highest value of CDW at 5.22g/L comprising 5% (w/v) of each carbon source and 3% (w/v) of each nitrogen source. In order to know which factor gave significant result in enhancing the CDW value, the analysis of variants (ANOVA) was made. From the ANOVA, the significant factors that contributed to the optimum growth of *P. aeruginosa* were factors C and D, palm oil, and ammonium nitrate, respectively.

The best combination of carbon and nitrogen sources occurred at the lowest value for both, suggesting the ability of the bacterial isolate to survive at low concentration of carbon and nitrogen. The ANOVA value for R-squared was 0.8969 and there was nonsignificant lack of fit; thus, this model was accepted.

13.4   CONCLUSIONS

The isolate was identified as *P. aeruginosa* PAO1 strain based on molecular sequencing. It has the potential in producing both PHAs and rhamnolipids according to FTIR analyses. Further study on optimization of both PHAs and rhamnolipids using response surface methodology (RSM) is suggested as both are very valuable in many fields of our industries such as environmental, pharmaceutical, agricultural, and also cosmetics.

The optimization study could be done using RSM as this method is more practical than conventional method in terms of time, cost of production and is faster to be analyzed. The Design Expert Software is recommended as a tool to conduct RSM during optimization.

13.5   ACKNOWLEDGMENT

The author would like to thank Research Management Institute (RMI) Universiti Teknologi Mara (UiTM) for funding this project and Institute of

Graduate Studies of UiTM. Special thanks to all staffs from Ionic, Colour and Coating Laboratory (ICC) and Microbiology Laboratory for their guidance and help in completing the screening process.

KEYWORDS

- polyhydroxyalkanoates
- *Pseudomonas aeruginosa*
- rhamnolipids
- biosurfactant
- full factorial design

REFERENCES

1. Anderson, A. J.; Dawes, E. A. Occurrence, Metabolism, Metabolic Role, and Industrial Uses of Bacterial PHAs. *Microbiol. Rev.* **1990,** *54,* 450–472.
2. Madison, L. L.; Huisman, G.W. Metabolic Engineering of Poly (3-hydroxyalkanoates): from DNA to Plastic. *Microbiol. Mol. Biol. Rev.* **1999,** *63,* 21–53.
3. Lee, H. J.; Rho, J. K.; Noghabi, K. A.; Lee, S. E.; Choi, M. H.; Yoon, S. C. Channeling of Intermediates Derived from Medium-chain Fatty Acids and de novo Synthesized Fatty Acids to Polyhydroxyalkanoic Acid by 2-bromoctanoic acid in *Pseudomonas fluorescens* BM07. *J. Microbiol. Biotechnol.* **2004,** *14,* 1256–1266.
4. Choi, M. H.; Xu, J.; Gutierrez, M.; Yoo, T.; Cho, Y. H.; Yoon, S. C. Metabolic Relationship between Polyhydroxyalkanoic Acid and Rhamnolipid Synthesis in *Pseudomonas aeruginosa*: Comparative ^{13}C NMR Analysis of the Products in Wild-type and Mutants. *J. Biotechnol.* **2011,** *151,* 30–42.
5. Reddy, C. S.; Ghai, R.; Rashmi; Kalia, V. C. PHAs: an Overview. *Bioresour. Technol.* **2003,** *87,* 137–146.
6. Soberón-Chávez, G.; Aguirre-Ramírez, M.; Sánchez, R. Production of Rhamnolipids by *Pseudomonas aeruginosa. Appl. Microbiol. Biotechnol.* **2005,** *68,* 718–725.
7. Hori, K.; Ichinohe, R.; Unno, H.; Marsudi, S. Simultaneous Syntheses of Polyhydroxyalkanoates and Rhamnolipids by *Pseudomonas aeruginosa* IFO3924 at Various Temperature and from Various Fatty Acids. *Biochem. Eng.* **2011,** *53,* 196–202.
8. Hori, K.; Marsudi, S.; Unno, H. Simultaneous Production of PHAs and Rhamnolipids by *Pseudomonas aeruginosa. Biotechnol. Bioeng.* **2002,** *78,* 699–707.
9. Marsudi, S.; Unno, H.; Hori, K. Palm Oil Utilization for the Simultaneous Production of PHAs and Rhamnolipids by *Pseudomonas aeruginosa. Appl. Microbiol. Biotechnol.* **2008,** *78,* 955–961.

10. Buchs, J. Introduction to Advantages and Problems of Shaken Cultures. *Biochem. Eng. J.* **2001**, *7*(2), 91–98.

11. Maier, U.; Buchs, J. Characterisation of the Gas-Liquid Mass Transfer in Shaking Bioreactors. *Biochem. Eng. J.* **2001**, *7*(2), 99–106.

12. Mortzek, C.; Anderlei, T.; Henzler, H.; Buchs, J. Mass Transfer Resistance of Sterile Plugs in Shaking Bioreactors. *Biochem. Eng. J.* **2001**, *7*, 107–112.

13. Stockmann, C.; Losen, M.; Dahlems, U.; Knocke, C.; Gellissen, G.; Buchs, J. Effect of Oxygen Supply on Passaging, Stabilising and Screening of Recombinant *Hansenula polymorpha* Production Strains in Test Tube Cultures. *FEMS Yeast Res.* **2003**, *4*, 195–205.

14. Bodour, A. A.; Dress, K. P.; Maier, R. M. Distribution of Biosurfactant-producing Bacteria in Undisturbed and Contaminated Arid Southwestern Soils. *Appl. Environ. Microbiol.* **2003**, *69*(6), 3280–3287.

15. Anandaraj, B.; Thivakaran, P. Isolation And Production Of Biosurfactant Producing Organism From Oil Spilled Soil. *J. Biosci Tech.* **2010**, *1*(3), 120–126.

16. Cooper, D. G.; Goldenberg, B. G. Surface-active Agents from Two Bacillus Species. *Appl. Environ. Microbiol.* **1987**, *53*, 224.

17. Liu, H. H.; Schmidt, S.; Poulsen, H.F.; Godfrey, A.; Liu, Z. Q.; Sharon, J. A; Huang, X. Three-Dimensional Orientation Mapping in the Transmission Electron Microscope. *Science* **2011**, *332*(6031), 833–834.

18. Pornsunthorntawee, N.; Arttaweeporn, N.; Paisanjit, S.; Somboonthanate, P.; Abe, M.; Rujiravanit, R.; Chavadej, S. Isolation and Comparison of Biosurfactants Produced by *Bacillus subtilis* PT2 and *Pseudomonas aeruginosa* SP4 for Microbial Surfactant-enhanced Oil Recovery. *Biochem. Eng. J.* **2008**, *42*, 172–179.

19. Ainon, H.; Noramiza, S.; Shahida, R. Screening and Optimization of Biosurfactant Production by the Hydrocarbon-degrading Bacteria. Sains Malaysiana **2013**, *42*, 5615–5623.

20. Campos-García, J.; Caro, A.D.; Nájera, R.; Miller-Maier, R. M.; Al-Tahhan, R.A.; Soberón-Chávez, G. The *Pseudomonas aeruginosa* rhlG Gene Encodes an NADPH-dependent β-ketoacyl Reductase Which is Specifically Involved in Rhamnolipid Synthesis. *J. Bacteriol.* **1998**, *180*, 4442–4451.

21. Christovaa, N.; Tulevaa, B.; Lalchevb, Z.; Jordanovac, A.; Jordanovd, B.; Naturforsch, Z. Rhamnolipid Biosurfactants Produced by *Renibacterium salmoninarum 27BN* During Growth on n-Hexadecane. *Zeitschrift fur Naturforschung C* **2004**, *59*(1–2), 70–74.

CHAPTER 14

SURVIVABILITY CHARACTERISTICS OF *BIFIDOBACTERIUM* SPP. ISOLATES FROM NEWBORN MECONIUM AND BREAST-FED/ FORMULATED INFANT FECES IN ACIDIC-SIMULATED INTESTINAL CONDITIONS

YUSHA FARZINI, MAT NOR ROHANA, and
ABDUL KHALIL KHALILAH*

Department of Biomolecular Sciences, Faculty of Applied Sciences, University of Technology MARA, Shah Alam, 40000, Selangor, Malaysia

Corresponding author. E-mail: khali552@salam.uitm.edu.my

CONTENTS

ABSTRACT

Although bifidobacteria has been suggested to be beneficial for the host and is component of many probiotics and competitive exclusion mixtures, the knowledge on abundance and their probiotic characteristic such as survivability in simulated intestinal fluids in isolates from local breast-fed and formulated infant feces is still limited. The present study was aimed to isolate *Bifidobacterium* strains from breast-fed and formulated infant feces and characterized the isolates based on their morphology, catalase activity, and survivability in acidic as well as simulated gastric (SGF)/ simulated intestinal fluid (SIF). About 25 fecal samples were collected and all isolates were grown for 48 hours, anaerobically in Man Rogosa and Sharp (MRS) with L-cystein (pH 6.4). Only isolates with rod and Y-shape morphology and catalase negative were selected for survivability test in acidic environments (pH 1, 2, and 3) as well as for survivability in intestinal and gastrointestinal fluids (SGF and SIF). From 68 isolates, 35 isolates were chosen based on their microscopic morphology. Out of 35 isolates, 33 isolates were found to be catalase negative. Among all 33 isolates selected, 17 isolates were shown a promising survivability in acidic environment as well as in SGF and SIF. Those 17 isolates were selected for further work on probiotic characteristics and molecular identification.

14.1 INTRODUCTION

It has been widely known that in the human gastrointestinal tract populations of various types of lactic acid bacteria thrive and some of which found to have probiotic characteristics. Types of bacteria such as *Lactobacillus* spp., *Bifidobacterium* spp., *Streptococci* spp., *Escherichia coli*, and various Enterobacteria were able to be obtained from the human gastrointestinal tract [1, 2]. As this study is focused on *Bifidobacterium* spp. strains, samples were obtained from infants ranging from 0 months to 12 months of age. *Bifidobacterium* spp. can be most easily found in gastrointestinal tract of infants during the earlier years of birth. It has been reported that as a child grows and is fed with mixed diet with solid food and formula milk, the cell population of *Bifidobacterium* spp. starts to reduce [3]. Breast-fed infants tend to have more complex populations of *Bifidobacterium* spp. compared with formula-fed infants [4–6].

Probiotics are defined as living microbial feed supplements consumed for health benefits, beyond providing basic nutritional value [7]. The use of probiotic throughout the world has grown largely significant due to its health benefits. Probiotic has been found to have positive effects on various types of diseases such as irritable bowel syndrome (IBS) [8], allergic rhinitis [9], hepatic damage [10], and antibiotic-associated diarrhea [11]. A study by Kanamori et al. [12] showed promising effects of probiotic treatment on infants with severe congenital anomaly and the strains used allowed activation of absorptive functions and prevented severe infections.

Probiotic bacteria especially *Lactobacillus* spp. and *Bifidobacterium* spp. have been widely used in food as well as in pharmaceutical industries. However, the use of probiotic over the years has also increased concern of possible risks they may have for humans, especially for neonates born prematurely or who have immune deficiency [13]. Thus, the search for safer and better probiotics is still extensively undergoing.

Previous research has succeeded in isolating probiotics from human gastrointestinal tract of people from various regions such as European countries, Japan, and some Asian countries [2, 14–16]. There are some lactic acid and probiotics studies that have been done in Malaysia which are focused on strains isolated from fermented food, breast milk, and animals [17–19]. However, studies on *Bifidobacterium* spp. isolated from our local infant feces either from breast-fed or formula-fed infants as well as from newborn meconium are still scarce.

Thus, the aim of this study is to isolate and study the probiotic characteristics of *Bifidobacterium* strains from local newborn meconium and breast-fed/formulated infant feces. The strains were selected based on their morphology, catalase activity, and their survivability in acidic environment as well as in simulated gastric/intestinal fluids. Low pH survivability was performed to screen for isolates which were able to survive at pH 1, 2, and 3. This is important to determine the isolates reliability in food manufacturing and also as a comparative factor for the following tests. Further tests include survivability in simulated gastrointestinal fluid (SGF) and simulated intestinal fluids (SIF) to determine their resistance to stomach juices. This is essential for the probiotic gastrointestinal delivery to the target region prior to their health benefit effects on the host.

14.2 MATERIALS AND METHODS

14.2.1 SUBJECTS OF SAMPLING

A total of five meconium samples were collected from healthy newborns, while 20 fecal samples (from breast-fed, formula-fed and mixed breast-fed–formulated infants) were collected and the age of infants ranged from 0 to 12 months. All samples were collected with consent from the parents, or the facilities which the parents entrusted care of their children. After collection, all fecal samples were stored in an icebox and transported to the laboratory within 48 h prior to isolation of pure cultures.

14.2.2 ISOLATION OF BIFIDOBACTERIUM

All samples were grown in de Man, Rogosa, and Sharp medium (MRS) broth supplemented with L-cysteine (0.5 g/L). They were incubated for 48 h in anaerobic conditions. Samples were then streaked onto MRS agar to obtain single isolated colonies. Distinguished colonies were selected and investigated microscopically and macroscopically, purified, and stored in MRS broth added with glycerol (40% w/v) and kept at −20°C for long-term storage. *B. animalis* and *B. infantis* strains were used as representative strains of *Bifidobacterium*.

14.2.3 CATALASE ACTIVITY BY H_2O_2 RESISTANCE

Each sample was tested for hydrogen peroxide (H_2O_2) resistance which indicates its capability of producing catalase. A drop of 3% hydrogen peroxide solution was applied onto a fresh colony which was placed inside a sterile Petri dish to limit aerosols which may be produced as catalase reacts with hydrogen peroxide. Presence of bubbles (O_2 + water resulted in formation of bubbles) indicates catalase positive isolate.

14.2.4 RESISTANCE TO LOW PH AND SURVIVABILITY TO SGF AND SIF JUICES

Acid tolerance: Fresh culture of 1 mL was transferred into three different tubes containing MRS broth of 9 mL volume and each tube was adjusted to initial pH of 1, 2 and 3, respectively). After inoculation, all tubes were incubated at 37°C for 4 h. Sampling was done at 0 and 4 h of incubation by plating onto MRS agar. The plates were further incubated for 48 h at 37°C anaerobically.

Survivability in SGF and SIF: Survival of cultures in SGF and SIF was determined using modified methods described by Bao et al. [20] and Nawaz et al. [16]. SGF (pH 2.5) was inoculated with 10% (v/v) fresh culture and incubated at 37°C under anaerobic condition. Subsequently, 1 mL of the cell suspension was further added into SIF (pH 8) and further incubation at 37°C under anaerobic conditions was carried out. Simulation of 100 uL volume were later plated using spread plate method on MRS agar after 0, 1, and 2 h incubation in SGF and after 0, 2, 4, and 6 h incubation in SIF. The plates were further incubated for another 48 h at 37°C under anaerobic conditions. All results were represented in $\log_{10}$ cfu/mL unit.

14.2.5 STATISTICAL ANALYSIS

The statistical analysis was performed using MINITAB version 14 (Minitab Inc., PA, United States). One-way ANOVA was used to examine significant differences between the normally distributed data. The mean values and the standard deviation were calculated from the data obtained through triplicate trials. A probability of $p < 0.05$ was used as a criterion for statistical significance.

14.3 RESULTS AND DISCUSSION

14.3.1 ISOLATION OF BIFIDOBACTERIUM

Among the 25 samples of newborn meconium and infant feces, 68 isolates were obtained. Different species of *Bifidobacterium* vary substantially in size and shape. The cell morphology with V-shaped, Y-shaped, rods, clubbed rods, or irregular-shaped cells of *Bifidobacterium* strains were observed. Typical colony morphology of *Bifidobacterium* spp. was observed as opaque, white color, and mostly convex and round, except for *B. adolescentis* and *B. longum* [21]. To reduce the number of isolates for further investigation, the cells that showed Gram positive and rod/Y-shape were selected, leaving only 35 isolates. It was noted that the number of potential isolates were found to be higher from fully breast-fed infants and mixed breast-fed formula-fed infants compared with formula-fed infants.

14.3.2   CATALASE ACTIVITY

The isolates were further tested for the presence of catalase by using H_2O_2. Only catalase-negative isolates were selected. Out of the 35 isolates, 33 showed to be catalase negative. Catalase test is one of the tools to differentiate between aerotolerant positive organisms. Strict anaerobes are mostly considered to be catalase negative [22].

14.3.3   SURVIVABILITY IN ACIDIC CONDITIONS

All 33 isolates were exposed to acidified media at pH 1, 2, and 3 for 4 h separately (Table 14.1). After 4 h of exposure to pH 1 condition, two isolates were found to show low rate of survivability. Meanwhile, another 20 isolates showed good survivability ranging from 2 to 4 $\log_{10}$ cfu/mL cells reduction from initial dose after the cells were exposed to pH 2 for 4 h. In contrast, all isolates were observed to survive at pH 3 after 4 h of exposure to acidic condition with cell reduction ranging from 0 to 2 $\log_{10}$ cfu/mL except for Pia, Sic, and Siia isolates which presented reduction of cell number ranging from 3 to 4 $\log_{10}$ cfu/mL from initial inoculum dose at 0 h. This result paralleled with previous findings by Shuhaimi et al. [23] who reported that most of the bifidobacteria isolates have low survivability at pH 1 and 2.

14.3.4   SURVIVABILITY IN SGF AND SIF

Table 14.2 shows the survivability of all 33 isolates in SGF and SIF. Reduction in the cell number in all isolates was observed with time of exposure. After all isolates were exposed for 2 h in SGF at pH 2.5, it was observed that almost all isolates were able to survive with cell reduction ranging from 1 to 3 $\log_{10}$ cfu/mL from the initial inoculum dose at 0 h of incubation. Meanwhile, only three isolates, namely, Pia, Sid, and Uiia were unable to survive after 2 h of exposure in SGF. This gave a promising indication that most isolates obtained were able to survive in extreme

TABLE 14.1 Acid Tolerance of Selected Isolates at Different Acidic Conditions.

Isolate	Survivability in Acidic Condition (log$_{10}$ cfu/mL)					
	pH 1		pH 2		pH 3	
	0 h	4 h	0 h	4 h	0 h	4 h
Dib	*5.95[a]**	0[a]	7.09[a]	4.10 ± 0.02[a]	6.77[a]	5.63 ± 0.04[a]
Gia	0[b]	0[a]	3.00[b]	0[b]	3.00[b]	2.61 ± 0.01[b]
Hia	0[b]	0[a]	5.92[c]	0.89 ± 0.05[c]	4.97[c]	4.02 ± 0.03[c]
Iiia	1.64[c]	0[a]	4.70[d]	0[b]	6.72[a]	6.62 ± 0.02[d]
Jia	2.51[d]	0[a]	5.61[c]	2.03 ± 0.60[d]	5.73[d]	4.56 ± 0.02[c]
Niia	0.85[e]	0[a]	6.50[a]	1.05 ± 0.05[c]	6.77[a]	7.08 ± 0.05[e]
Oia	3.08[f]	0[a]	5.63[c]	0[b]	5.66[d]	4.12 ± 0.01[c]
Pia	0[b]	0[a]	6.18[a]	0[b]	6.20[a]	2.76 ± 0.02[b]
Ric	0.70[e]	0[a]	2.30[e]	0[b]	2.48[b]	2.64 ± 0.02[b]
Riia	0[b]	0[a]	6.40[a]	0[b]	6.58[a]	6.09 ± 0.01[a]
Riib	0[b]	0[a]	6.51[a]	3.07 ± 0.01[e]	6.55[a]	6.23 ± 0.01[d]
Sib	0[b]	0[a]	5.70[c]	0.80 ± 0.03[c]	6.38[a]	6.46 ± 0.02[d]
Sic	0[b]	0[a]	6.50[a]	0[b]	6.85[a]	3.42 ± 0.03[f]
Sid	0[b]	0[a]	6.83[a]	4.18 ± 0.01[a]	5.78[d]	7.18 ± 0.01[e]
Siia	0[b]	0[a]	5.40[c]	0[b]	5.70[d]	2.25 ± 0.00[b]
Siib	0[b]	0[a]	5.65[c]	1.68 ± 0.10	6.08[d]	5.54 ± 0.09[a]
Tia	1.72[c]	0[a]	5.45[c]	0[b]	5.30[d]	5.00 ± 0.00[c]
Tib	0[b]	0[a]	5.08[d]	0[b]	4.16[c]	5.93 ± 0.04[a]
Tic	0.30[g]	0[a]	6.47[a]	0[b]	6.45[a]	5.35 ± 0.50[a]

TABLE 14.1 (Continued)

Isolate	Survivability in Acidic Condition (log₁₀ cfu/mL)					
	pH 1		pH 2		pH 3	
	0 h	4 h	0 h	4 h	0 h	4 h
Tiia	2.28[d]	0[a]	6.66[a]	0[b]	6.76[a]	6.24 ± 0.09[d]
Uia	0[b]	0[a]	5.23[c]	0[b]	5.48	5.87 ± 0.24[a]
Uib	2.70[d]	0[a]	5.99[c]	0[b]	6.35[a]	4.20 ± 0.10[c]
Uiia	0[b]	0[a]	5.60[c]	0[b]	4.95[d]	5.00 ± 0.00[c]
Via	2.68[d]	0[a]	5.75[c]	0.95 ± 0.02[c]	5.53[d]	5.54 ± 0.09[a]
Viia	2.58[d]	0[a]	6.05[c]	0.82 ± 0.06[c]	5.89[d]	6.80 ± 0.01[d]
Viib	2.80[d]	0[a]	5.90[c]	0.85 ± 0.09[c]	6.55[a]	7.01 ± 0.05[c]
Wia	1.54[c]	0.90 ± 0.60[b]	6.32[a]	0.56 ± 0.79[c]	6.59[a]	5.39 ± 0.12[a]
Wiia	0[b]	0[a]	6.56[a]	1.02 ± 0.01[d]	6.51[a]	5.54 ± 0.09[a]
Xia	4.13[h]	0[a]	5.77[c]	4.19 ± 0.02[a]	6.77[a]	6.68 ± 0.53[d]
Yaxi	3.88[i]	0[a]	6.23[a]	3.21 ± 0.04[e]	6.98[a]	6.99 ± 0.21[d]
Ybxii	3.15[f]	0.27 ± 0.09[c]	6.70[a]	4.80 ± 0.09[a]	6.91[a]	6.23 ± 0.08[d]
Yei	3.60[i]	0[a]	6.84[a]	4.56 ± 0.11[a]	6.75[a]	6.02 ± 0.02[a]
Yexi	2.98[f]	0[a]	5.22[c]	3.26 ± 0.02[e]	6.95[a]	5.65 ± 0.02[a]
B. animalis	3.09[f]	0[a]	6.07[c]	4.34 ± 0.05[a]	6.41[a]	6.02 ± 0.01[a]
B. infantis	3.01[f]	0[a]	6.49[a]	4.80 ± 0.10[a]	6.56[a]	6.29 ± 0.03[d]

[a]Values were obtained based on the duplicate experiments.
[b]Different small letters in the same column indicate significant difference ($p < 0.05$).
B. animalis and B. infantis are representative strains of Bifidobacterium.

TABLE 14.2 Survivability in Simulated Gastric (SGF) and Intestinal Fluids (SIF) at pH 2.5 and 8, respectively (log10 cfu/mL).

| Isolate | Survivability in Gastrointestinal (SGF) and Intestinal Fluids (SIF) (log cfu/mL) | | | | | | |
| | SGF (pH 2.5) | | | SIF (pH 8) | | | |
	0 h	1 h	2 h	0 h	2 h	4 h	6 h
Dib	*5.96^a**	4.17 ± 0.00^a	4.15 ± 0.01^a	2.74 ± 0.00^a	2.66 ± 0.03^a	2.62 ± 0.01^a	2.60 ± 0.00^a
Gia	5.85^a	4.15 ± 0.02^a	3.57 ± 0.01^b	0.64 ± 0.20^b	0.12 ± 0.17^b	0^b	0^b
Hia	5.60^a	3.52 ± 0.02^b	3.12 ± 0.05^b	2.01 ± 0.00^c	1.00 ± 0.01^c	0.88 ± 0.01^c	0.70 ± 0.04^c
Iiia	5.10^a	3.67 ± 0.01^b	3.54 ± 0.00^b	0.84 ± 0.09^b	0.30 ± 0.43^b	0.30 ± 0.00^c	0^b
Jia	6.78^b	5.33 ± 0.41^c	5.31 ± 0.02^c	3.79 ± 0.00^d	3.63 ± 0.01^d	3.03 ± 0.01^d	2.98 ± 0.00^a
Niia	6.97^b	4.83 ± 0.07^a	3.99 ± 0.12^b	2.21 ± 0.01^a	1.88 ± 0.00^c	1.72 ± 0.02^c	1.67 ± 0.01^d
Oia	5.78	3.60 ± 0.02^b	3.15 ± 0.08^b	0^f	0^e	0^b	0^b
Pia	5.98^a	0^d	0^d	0^f	0^e	0^b	0^b
Ric	5.70^a	5.70 ± 0.01^c	4.06 ± 0.02^a	1.35 ± 0.01^c	0.87 ± 0.04^b	0^b	0^b
Riia	6.79^b	5.54 ± 0.08^c	4.58 ± 0.04^a	2.80 ± 0.01^a	1.35 ± 0.01^c	0.20 ± 0.06^c	0^b
Riib	5.66^a	4.81 ± 0.08^a	4.77 ± 0.01^a	1.74 ± 0.09^e	1.71 ± 0.04^c	1.44 ± 0.11^e	1.37 ± 0.01^d
Sib	8.75	6.42 ± 0.00^e	5.61 ± 0.02^c	3.75 ± 0.01^d	1.62 ± 0.18^c	0.69 ± 0.12^c	0.15 ± 0.21^c
Sic	5.24^a	4.04 ± 0.09^a	3.73 ± 0.03^b	1.48 ± 0.03^c	0.62 ± 0.23^c	0^b	0^b
Sid	5.20^a	3.73 ± 0.16^b	0^d	0^f	0^e	0^b	0^b
Siia	5.63^a	4.49 ± 0.02^a	3.51 ± 0.21^b	0.93 ± 0.36^g	0.35 ± 0.50^b	0^b	0^b
Siib	5.45^a	4.33 ± 0.18^a	3.88 ± 0.04^b	2.60 ± 0.01^a	2.49 ± 0.01^a	2.48 ± 0.01^a	2.45 ± 0.01^a
Tia	5.85^a	4.57 ± 0.03^a	4.26 ± 0.04^a	3.02 ± 0.21^d	1.10 ± 0.11^c	0^b	0^b
Tib	6.72^b	5.62 ± 0.01^c	5.28 ± 0.01^c	2.74 ± 0.02^a	1.12 ± 0.10^c	0^b	0^b
Tic	5.74^a	3.67 ± 0.01^c	2.30 ± 0.01^c	0.78 ± 0.25^g	0^e	0^b	0^b

TABLE 14.2 (Continued)

| Isolate | Survivability in Gastrointestinal (SGF) and Intestinal Fluids (SIF) (log cfu/mL) | | | | | | |
| | SGF (pH 2.5) | | | SIF (pH 8) | | | |
	0 h	1 h	2 h	0 h	2 h	4 h	6 h
Tiia	6.30[b]	6.11 ± 0.10[e]	5.19 ± 0.02[c]	2.51 ± 0.01[a]	0.15 ± 0.20[b]	0[b]	0[b]
Uia	6.08	5.23 ± 0.22[c]	3.04 ± 0.00[b]	1.36 ± 0.12[e]	0.24 ± 0.34[b]	0[b]	0[b]
Uib	5.68	3.42 ± 0.04[b]	2.03 ± 0.01[e]	0.15 ± 0.21[f]	0[e]	0[b]	0[b]
Uiia	5.57[a]	2.48 ± 0.01[b]	0[d]	0[f]	0[e]	0[b]	0[b]
Via	5.75[a]	4.35 ± 0.06[a]	4.18 ± 0.02[a]	2.24 ± 0.01[c]	2.02 ± 0.09[a]	1.80 ± 0.03[e]	1.42 ± 0.07[d]
Viia	7.08[b]	6.04 ± 0.01[e]	4.71 ± 0.00[a]	3.41 ± 0.01[d]	3.34 ± 0.01[d]	3.24 ± 0.01[d]	2.95 ± 0.01[a]
Vib	6.93[b]	5.95 ± 0.02[c]	5.72 ± 0.02[c]	4.47 ± 0.01[a]	4.12 ± 0.02[f]	4.08 ± 0.01[d]	4.07 ± 0.00[e]
Wia	6.48[b]	6.43 ± 0.02[e]	5.37 ± 0.01[c]	4.24 ± 0.01[a]	4.18 ± 0.01[f]	4.17 ± 0.00[d]	4.12 ± 0.01[e]
Wiia	6.11[b]	6.10 ± 0.02[e]	5.10 ± 0.01[c]	4.19 ± 0.01[a]	4.16 ± 0.01[f]	4.15 ± 0.01[d]	4.12 ± 0.01[e]
Xia	5.46[a]	5.20 ± 0.05[e]	4.69 ± 0.01[a]	2.34 ± 0.19[a]	2.06 ± 0.02[a]	1.46 ± 0.19[e]	0.44 ± 0.19[e]
Yaxi	6.95[b]	6.89 ± 0.15[e]	6.39 ± 0.03[f]	5.17 ± 0.03[h]	4.74 ± 0.01[f]	4.62 ± 0.01[d]	4.54 ± 0.02[e]
Ybxii	8.80	5.81 ± 0.05[c]	5.68 ± 0.04[c]	4.94 ± 0.01[a]	4.72 ± 0.01[f]	3.74 ± 0.02[d]	3.49 ± 0.01[f]
Yei	5.74[a]	4.80 ± 0.02[a]	2.76 ± 0.03[e]	2.39 ± 0.12[c]	1.92 ± 0.05[c]	1.43 ± 0.07[e]	0.98 ± 0.03[e]
Yexi	7.26	7.14 ± 0.09[f]	6.11 ± 0.02[f]	5.04 ± 0.03[h]	4.81 ± 0.01[f]	4.41 ± 0.05[d]	4.31 ± 0.05[e]
B. animalis	7.79	6.32 ± 0.02[e]	4.88 ± 0.05[a]	3.85 ± 0.03[d]	3.13 ± 0.02[d]	2.90 ± 0.03[a]	2.83 ± 0.01[a]
B. infantis	7.70	6.09 ± 0.01[c]	5.46 ± 0.01[c]	3.48 ± 0.01[d]	2.88 ± 0.01[a]	2.44 ± 0.02[a]	2.37 ± 0.02[a]

*Values were obtained based on the duplicate experiments.
**Different small letters in the same column indicate significant difference ($p < 0.05$).
B. animalis and B. infantis are representative strains of Bifidobacterium.

gastric condition and this is one of the important criteria in probiotic strain selection. Meanwhile in SIF exposure, similar scenario was observed with cell survival reduced with time of exposure. The number of cell loss increased after 2, 4, and 6 h of SIF exposure to pH 8 at 18%, 42%, and 48%, respectively.

Based on these results, almost all *Bifidobacterium* isolates were able to survive both at acidic condition of pH 2 and at alkaline condition of pH 8 in SGF and SIF, respectively. About 17 *Bifidobacterium* isolates, namely, Jia, Niia, Riib, Siib, Viia, Viib, Wia, Wiia, Xia, Yaxi, Ybxii, Yei, and Yexi showed a promising survivability in acidic conditions as well as in SGF–SIF in which the results obtained were comparable to the ones produced by the comparative strains used in this study.

14.4 CONCLUSION

In conclusion, 17 isolates showed good survivability in all of the tests done. These isolates will be further studied based on other probiotic characteristics such as bile salt tolerance, antibiotic resistance, antimicrobial activity, and molecular identification using 16S rRNA.

KEYWORDS

- isolation
- *Bifidobacterium*
- meconium
- feces
- infant survivability
- gastrointestinal

REFERENCES

1. Savage, D. C. Microbial Biota of the Human Intestine: A Tribute to Some Pioneering Scientists. *Curr. Issues Intest. Microbiol.* **2001,** *2*(1), 1–15.

2. Reuter, G. The Lactobacillus and Bifidobacterium Microflora of the Human Intestine: Composition and Succession. *Curr. Issues Intest. Microbiol.* **2001**, *2*(2), 43–53.

3. Sun, Z.; Baur, A.; Zhurina, D.; Yuan, J.; Christian, U. R. Accessing the Inaccessible: Molecular Tools for Bifidobacteria. *Appl. Environ. Microbiol.* **2012**, *78*(5), 5035–5042.

4. Roger, L. C.; Costabile, A.; Holland, D. T.; Hoyles, L.; McCartney, A. L. Examination of Faecal Bifidobacterium Populations in Breast- and Formula-fed Infants during the First 18 Months of Life. *Microbiology* **2010**, *156*, 3329–3341.

5. Klaassens, E. S.; Boesten, R. J.; Haarman, M.; Knol, J.; Schuren, F. H.; Vaughan, E. E.; de Vos, W. M. Mixed-species Genomic Microarray Analysis of Fecal Samples Reveals Differential Transcriptional Responses of Bifidobacteria in Breast- and Formula-Fed Infants. *Appl. Environ. Microbiol.* **2009**, *75*, 2668–2676.

6. Turroni, F.; Peano, C.; Pass, D. A.; Foroni, E.; Severgnini, M.; Claesson, M. J.; Kerr, C.; Hourihane, J.; Murray, D.; Fuligni, F.; Gueimonde, M.; Margolles, A.; De Bellis, G.; O'Toole, P. W.; Sinderen, D.; Marchesi, J. R.; Venture, M. Diversity of Bifidobacteria within the Infant Gut Microbiota. *PLoS ONE* **2012**, *7*(5), e36957.

7. Parvez, S.; Malik, K. A.; Kang, S. A.; Kim, H.-Y. Probiotics and Their Fermented Food Products Are Beneficial for Health. *J. App. Microbiol.* **2006**, *100*, 1171–1185.

8. Hedin, C, R, H.; Mullard, M.; Sharratt, E.; Jansen, C.; Sanderson, J.; Shirlaw, P.; Howe, L. C.; Djemal, S.; Stagg, A, J.; Lindsay, J, O.; Whelan, K. Probiotic and Probiotic Use in Patients with Inflammatory Bowel Disease: A Case–Control Study. *Inflam. Bowel Dis.* **2010**, *16*(12), 2099–2108.

9. Ivory, K.; Chambers, S. J.; Pin, C.; Prieto, E.; Arqués, J. L.; Nicoletti, C. Oral Delivery of *Lactobacillus casei* Shirota Modifies Allergen-induced Immune Responses in Allergic Rhinitis. *Clin. Exp. Allergy* **2008**, *38*, 1282–1289.

10. Ewaschuk, J.; Endersby, R.; Thiel D.; Diaz, H.; Backer, J.; Ma, M.; Churchill, T.; Madsen, K. Probiotic Bacteria Prevent Hepatic Damage and Maintain Colonic Barrier Function in a Mouse Model of Sepsis. *Hepatology* **2007**, *46*, 841–850.

11. Wenus, C.; Goll, R.; Loken, E. B.; Biong, A. S.; Halvorsen, D. S.; Florholmen, J. Prevention of Antibiotic-associated Diarrhoea by Fermented Probiotic Milk Drink. *Eur. J. Clin. Nutr.* **2008**, *62*, 299–301.

12. Kanamori, Y.; Iwanaka, T.; Sugiyama, M.; Komura, M.; Takahashi, T.; Yuki, N.; Morotomi, M.; Tanaka, R. Early Use of Probiotics Is Important Therapy in Infants with Severe Congenital Anomaly. *Pediatr. Int.* **2010**, *52*, 362–367.

13. Boyle, R. J.; Robins-Browne, R. M.; Tang, M. L. K. Probiotic Use in Clinical Practice: What Are the Risks? *Am. J. Clin. Nutr.* **2006**, *83*, 1256–1264.

14. Dunne, C.; O'Mahony, L.; Murphy, L.; Thornton, G.; Morrissey O'Halloran, S.; Feeney, M.; Flynn, S.; Fitzgerald, G.; Daly, C.; Kiely, B.; O'Sullivan, G. C.; Shahanan, F.; Collins, J. K. In Vitro Selection Criteria for Probiotic Bacteria of Human Origin: Correlation with In Vivo Findings. *Am. J. Clin. Nutr.* **2001**, *73*(suppl), 386S–392S.

15. Morelli, L. In Vitro Selection of Probiotic Lactobacilli: A Critical Appraisal. *Curr. Issues Intest. Microbiol.* **2000**, *1*(2), 59–67.

16. Nawaz, M.; Wang, J.; Zhou, A.; Ma, C.; Wu, X.; Xu, J. Screening and Characterization of New Potentially Probiotic Lactobacilli from Breast-fed Healthy Babies in Pakistan. *Afr. J. Microbiol. Res.* **2011**, *5*(12), 1428–1436.

17. Belal, J. M.; Zaiton, H.; Mohamed, M. A. I.; Fredy, K. S. A. K.; Mohamad, M. A. Malaysian Isolates of Lactic Acid Bacteria with Antibacterial Activity against Gram-positive and Gram-negative Pathogenic Bacteria. *J. Food Res.* **2012,** *1*(1), 110–116.

18. Nurhidayu, A.; Ina-Salwany, M. Y.; Mohd Daud, H.; Harmin, S. A. Isolation, Screening, and Characterization of Potential Probiotics from Frarmed Tiger Grouper (*Epinephelus fuscoguttatus*). *Afr. J. Microbiol. Res.* **2011,** *6*(9), 1924–1933.

19. Roslinda, A. M.; Sallehhuddin, H.; Hesham, A. E. E.; Nor, Z. O.; Noor, A. Z.; Mohamad, R. S.; Ramlan, A. A. Production of *Lactobacillus salivarius*, a New Probiotic Strain Isolated from Human Breast Milk, in Semi-industrial Scale and Studies on Its Functional Characterizations. *Curr. Res. Technol. Educ. Top. Appl. Microbiol. Microb. Biotechnol.* **2010,** *11*, 1196–1204.

20. Bao, Y.; Zhang, Y.; Zhang, Y.; Liu, Y.; Wang, S.; Dong, X.; Wang, Y.; Zhang, H. Screening of Potential Probiotic Properties of *Lactobacillus fermentum* Isolated from Traditional Fermented Products. *Food Control* **2010,** *21*, 695–701.

21. Park, Y. P.; Lee, D. K.; An, H. M.; Cha, M. G.; Baek, E. H.; Kim, J. R.; Lee, S. W.; Kim, M. J.; Lee, K. O.; Ha, N. J. Phenotypic and Genotypic Characterization of *Bifidobacterium* Isolates from Healthy Adult Koreans. *Iran. J. Biotechnol.* **2011,** *9*(3), 173–180.

22. Martin, R.; Esther, J.; Hans, H.; Leonides, F.; Maria, L. M.; Erwin, G. Z.; Juan, M. R. Isolation of Bifidobacteria from Breast Milk and Assessment of Bifodobacterial Population by PCR-Denaturing Gradient Gel Electrophoresis and Quantitiative Real-Time PCR. *Appl. Environ. Microbiol.* **2009,** *75*, 965–969.

23. Shuhaimi, M.; Yazid, A. M.; Ali, A. M.; Ghazali, M. H.; Zaitun, H.; Nur Atiqah, N. A. Acid Adaptation of Bifidobacteria Isolated from Human Stools to Simulated pH of Human Stomach. *Pak. J. Biol. Sci.* **1999,** *2*(4), 1203–1206.

CHAPTER 15

ANTIOXIDANT AND ANTI-BACTERIAL POTENTIAL OF LICHEN SPECIES FROM MALAYSIA

LUQMAN SAIDI, MOHD FAIZ FOONG ABDULLAH*,
KAMSANI NGALIB, WAN YUNUS WAN AHMAD,
MOHD ROZI AHMAD, ASMIDA ISMAIL, and
MUHAMMAD ISMAIL ABD KADIR

*Faculty of Applied Sciences, Universiti Teknologi MARA, 40450
Shah Alam, Malaysia*

Corresponding author. E-mail: mohdf184@salam.uitm.edu.my

CONTENTS

ABSTRACT

The aim of this study is to investigate antioxidant and antibacterial activity of the methanol and acetone extracts of the lichens *Ramalina dumeticola, Usnea baileyi, Heterodermia leucomela,* and *Telochistes flavicans*. Total phenolic contents of the lichens are evaluated by Folin–Ciocalteu method. Both acetone and methanol extracts of *R. dumeticola* showed the highest total phenolic content which are 179.3 mg and 142.9 mg GAE/g of extract, respectively. Antioxidant activity was evaluated by DPPH free radical scavenging assay. Among the lichens tested, methanol extract of *R. dumeticola* showed the largest free radical scavenging activity (84.43% inhibition at a concentration of 1 mg/mL). The antibacterial activity was estimated by disc-diffusion method and also the minimal inhibitory concentration (MIC) by the broth microdilution method tested on six bacterial strains. *R. dumeticola* and *U. baileyi* extracts are the most active toward *B. subtilis, S. aureus,* and *S. epidermidis*. The present study showed that some of lichen extracts demonstrated a strong antioxidant and antibacterial effect.

15.1 INTRODUCTION

For past few decades, there was a great concern for discovery and development of new pharmacological active compounds from natural origins in the control and prevention of various human, animal, and plant diseases. It is known that long-term use of synthetic drugs often causes numerous side effects and sometimes develops resistance toward diseases [11]. Natural products are known to provide beneficial effects to the whole organism without adverse side effects. Due to their wide spectrum of known biological activities, lichens have become one of the most promising sources of drug discovery [12].

Lichens are obligate symbiotic systems consisting of a filamentous fungus and a photosynthetic partner (eukaryotic algae and/or cyanobacterium), and in some cases, nonphotosynthetic bacteria [8, 23]. Throughout the ages, lichens have been used for human and animal nutrition, coloring, perfumes, alcohol, and in the pharmaceutical industries [21, 14].

Lichens produce secondary metabolites called "lichen substances," which comprise depsides, depsidones, dibenzofurans, xanthones, and terpene derivatives [10]. These metabolites can constitute more than 30% of the lichen thallus dry mass [5]. These lichen metabolites are responsible for various biological activities such as antimicrobial, antiviral, antitumor, anti-inflammatory, analgesic, antipyretic, antiproliferative, and antiprotozoal [6].

Hence, the aim of this study is to investigate the species of lichens that have the potential to become a natural antioxidant and exhibit antibacterial activity against pathogenic bacteria that cause infection to animal and human.

15.2 MATERIALS AND METHODS

15.2.1 COLLECTION AND IDENTIFICATION OF LICHEN SPECIES

Lichens were collected at Cameron Highland and identified based on their morphology. The sorted lichens were dried at room temperature for a week.

15.2.2 PREPARATION OF LICHEN EXTRACT

Lichen samples were ground finely. A total of 20 g of the lichen powder was extracted using acetone and methanol separately (400 mL × 2) for 24 h by placing on an orbital shaker with 120 rpm at room temperature. The crude acetone and methanol extracts were filtered using Whatman® filter paper No 1. The solvent was removed using a rotary evaporator under reduced atmospheric pressure at 40°C. The dried extracts were stored in the freezer at −20°C for further analysis [26].

15.2.3 ANTIOXIDANT ACTIVITY

15.2.3.1 TOTAL PHENOLIC CONTENT

Total phenolic contents of all lichen extracts were determined using Folin–Ciocalteu reagent as described by Singleton and Rossi [24] and Maizura et al. [16]. Methanolic solution of the extract in the concentration of 1 mg/mL was used in the analysis. The reaction mixture was prepared by mixing 0.5 mL of methanolic solution of extract, 2.5 mL of 10% Folin–Ciocalteu reagent dissolved in water and 2.5 mL 7.5% Na_2CO_3. Blank was prepared, containing 0.5 mL methanol, 2.5 mL 10% Folin–Ciocalteu reagent dissolved in water and 2.5 mL of 7.5% of Na_2CO_3. The samples were incubated at room temperature for 2 h. The absorbance was determined using spectrophotometer at 765 nm. The samples were prepared in triplicate for each analysis and the mean value of absorbance was obtained. The standard curve of gallic acid solution (0.05, 0.1, 0.15, 0.25, 0.5 mg/mL) was prepared using the similar procedure and the standard calibration line was constructed. Total phenolic contents of the lichen extracts were expressed in terms of gallic acid equivalent (mg of GAE/g of extract).

15.2.3.2 DPPH RADICAL SCAVENGING ASSAY

DPPH free radical scavenging activity of the studied lichen extracts was estimated using modified method described by Blois [4]. About 2 mL of DPPH solution (0.1 mM of DPPH in methanol) was mixed with 1 mL (1 mg/mL in methanol) of the lichen extract. The mixtures were gently shaken, covered with aluminum foil, and incubated in the dark at room temperature (27°C) for 30 min. The absorbance of the resulting solution was measured at 517 nm against methanol as blank using a visible spectrophotometer to measure the content of remaining DPPH free radical. A solution of 2 mL of DPPH and 1 mL of methanol was used as a control. Ascorbic acid, butylated hydroxyanisole (BHA), and trolox were used as reference antioxidants for this test. The DPPH radical scavenging activity was calculated as percentage of DPPH• discoloration using the formula:

DPPH scavenging activity (%) = ((A0 − A1)/A0) × 100,

where A0 is the absorbance of the negative control and A1 is the absorbance of reaction mixture or standards.

15.2.4 ANTIBACTERIAL ACTIVITY

15.2.4.1 MICROORGANISMS

Staphylococcus aureus ATCC 6538, *Staphylococcus epidermidis* ATCC 12228, *Bacillus subtilis* ATCC 6633, *Escherichia coli* ATCC 25922, *Pseudomonas aeruginosa* ATCC 9027, and *Klebsiella pneumoniae* ATCC

4352 were used as test organisms in this study. Bacterial cultures were maintained on Mueller–Hinton agar substrates. All cultures were stored at 4°C and subcultured every 15 days.

15.2.4.2 DISC-DIFFUSION METHOD

A standard Kirby-Bauer disc-diffusion method was used to determine antimicrobial activity. Bacterial inoculate was obtained from bacterial cultures incubated for 24 h at 37°C in Mueller–Hinton broth and brought up by dilution according to the 0.5 McFarland standard to approximately 10^8 CFU/mL. The inoculums were swabbed onto Mueller–Hinton agar using sterile cotton swab. Sterile paper discs (6 mm diameter) were laid on the inoculated substrate after being pipetted with 20 µL of lichen extract (50 mg/mL). The bacterial cultures were incubated for 24 h at 37°C. Antimicrobial activity was determined by measuring the diameter of the inhibition zone around the disc. Streptomycin was used as a positive control. A DMSO solution was used as a negative control for the solvent influence. All experiments were performed in triplicate.

15.2.4.3 MINIMAL INHIBITORY CONCENTRATION

The minimal inhibitory concentration (MIC) was determined by the broth microdilution method using 96-well microtiter plates. A series of dilutions with concentrations ranging from 16 to 0.03125 mg/mL for extracts was used in the experiment against every microorganism tested. Twofold dilutions of extracts were prepared in Mueller–Hinton broth for bacterial cultures. The MIC of the extracts was determined using resazurin. A DMSO solution with inoculums was used as a negative control to determine the influence of the solvents. Broth with lichen extract was used as a positive control. All experiments were performed in triplicate.

15.2.5 STATISTICAL ANALYSIS

All statistical analysis was performed with the Microsoft EXCEL 2007. Pearson's bivariate correlation test was carried out to calculate correlation coefficients (r) between the total phenolic content and the DPPH radical scavenging activity. All data were expressed as mean ± SD from a minimum of three measurements.

15.3   RESULTS

15.3.1   ANTIOXIDANT ACTIVITY

15.3.1.1   TOTAL PHENOLIC CONTENT

Total phenolic content of both acetone and methanol extracts of the lichen species is shown in Figure 15.1. The amount of total phenolic contents was determined as the gallic acid equivalent using an equation obtained from standard gallic acid curve.

$$y = 8.1417x + 0.2787 \ (R^2 = 0.9955).$$

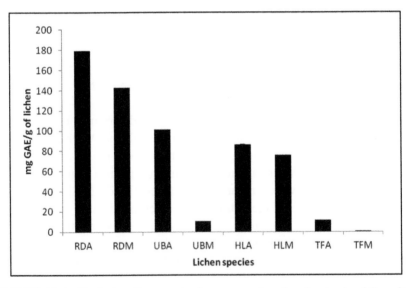

FIGURE 15.1   Total phenolic content of acetone and methanol extracts of *Ramalina dumeticola, Usnea baileyi, Heterodermia leucomela, and Teloschistes flavicans*. Symbols used for the extract are RDA—acetone extract of *R. dumeticola*, RDM—methanol extract of *R. dumeticola*, UBA—acetone extract of *U. baileyi*, and UBM—methanol extract of *U. baileyi*, HLA—acetone extract of *H. leucomela*, HLM—methanol extract of *H. leucomela*, TFA—acetone extract of *T. flavicans*, and TFM—methanol extract of *T. flavicans*. Values are mean ± SD (*n* = 3).

The highest total phenolic content was identified in the lichen *Ramalina dumeticola* in both acetone (179.3 ± 0.104) and methanol extract (142.9 ± 0.006) mg of GAE/g of lichen extract, respectively. Lichen of *Teloschistes flavicans* had shown the lowest total phenolic content in the methanol extract (0.18 ± 0.01) mg of GAE/g of extract and not visible on the bar chart, while the acetone extract had given phenolic content (11.2 ± 0.022) mg of GAE/g of extract. Both acetone and methanol extract of *Usnea baileyi* had shown significant differences in their total phenolic content which is higher in the acetone extract at (101 ± 0.011) and lower in the methanol extract (10.5 ± 0.007) mg GAE/g of extract, respectively. *Heterodermia leucomela* have almost the same amount of total phenolic content in both acetone and methanol extract, which are (86 ± 0.24) and (76 ± 0.021) mg GAE/g of extract, respectively.

15.3.1.2 DPPH RADICAL SCAVENGING ACTIVITY

Figure 15.2 shows the antioxidant activity evaluated by DPPH free radical scavenging assay of the studied lichen extracts of *R. dumeticola,*

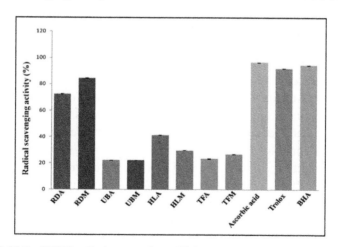

FIGURE 15.2 DPPH radical scavenging of lichen extracts and standard antioxidant; ascorbic acid, trolox, and BHA at concentration 1 mg/mL. Symbols used for the extract are RDA—acetone extract of *Ramalina dumeticola*, RDM—methanol extract of *Ramalina dumeticola*, UBA—acetone extract of *Usnea baileyi*, UBM—methanolic extract of *Usnea baileyi*, HLA—acetone extract of *Heterodermia leucomela*, HLM—methanol extracts of *Heterodermia leucomela*, TFA—acetone extract of *Teloschistes flavicans*, and TFM—methanol extract of *Teloschistes flavicans*. Values are mean ± SD (*n* = 3).

U. baileyi, H. leucomela, and *T. flavicans.* At a concentration of 1 mg/ mL, the methanol extract of *R. dumeticola* showed largest DPPH radical scavenging activity (84.43%) followed by the acetone extract (72.27%), which both nearly as standard antioxidants; ascorbic acid (96.45%), trolox (91.9%), and BHA (94.44%). Both acetone and methanol extracts of *U. baileyi* give similar radical scavenging activity (21.9%). The scavenging activity for the acetone extract of *H. leucomela* (41.17%) is higher than the methanol extract (29.61%). Both acetone and methanol extracts of *T. flavicans* give radical scavenging activity of 23.21% and 26.55%, respectively.

Most of the lichen extracts that exhibited the higher radical scavenging activity contain the high amount of total phenolic contents. The correlation coefficient between total phenolic content and free radical scavenging activity of the lichen extracts was $r = 0.811$.

15.4.1 ANTIBACTERIAL ACTIVITY

15.4.1.1 DISC-DIFFUSION METHOD

The antibacterial activity varies for each lichen extract and the tested organisms as shown in Table 15.1. Both acetone and methanol extracts of *R. dumeticola* exhibit a strong antibacterial effect toward *B. subtilis, S. aureus,* and *S. epidermidis* with a zone of inhibition ranging from 13 to 18 mm, but weak against *E. coli* and *K. pneumonia* with inhibition zone of 8 mm. *P. aeruginosa* are resistance toward both extracts.

Acetone and methanol extracts of *U. baileyi* showed strong antibacterial activities only toward *B. subtilis* and *S. epidermidis* with zone of inhibition ranging from 14 to 17 mm and weak on the others bacteria. Acetone extract had bacteriostatic effect on *S. epidermidis* (11 mm), *K. pneumonia,* and *P. aeruginosa* (both 19 mm).

The extracts of lichen *H. leucomela* exhibit weak antibacterial effects toward all tested bacteria with zone of inhibition below 10 mm. *T. flavicans* species of both acetone and methanol extracts showed strong antibacterial effect only toward *S. aureus* with zone of inhibition of 18 mm and 14 mm, respectively. The other tested bacteria showed weak effect of extract with zone of inhibition below 10 mm.

TABLE 15.1 Diameter of Inhibition Zone of *Ramalina dumeticola, Usnea baileyi, Heterodermia leucomela,* and *Teloschistes flavicans* Extracts Against the Test Organisms.

| Organisms | Diameter of Inhibition Zone (mm) | | | | | | | | |
| | *R. dumeticola* | | *U. baileyi* | | *H. leucomela* | | *T. flavicans* | | Anti-biotics |
	A	M	A	M	A	M	A	M	S
Bacillus subtilis	15^b	13	15	17	6	6	9	7	17
Staphylococcus aureus	16	15	9	11	9	8	18	14	14
Staphylococcus epidermidis	18	16	14	14	6	10	10	8	21
Escherichia coli	8	8	7	7	6	7	8	8	14
Klebsiella pneumoniae	8	8	7	19	8	8	8	8	15
Pseudomonas aeruginosa	6	6	6	19	6	6	11	6	13

A—acetone extract; M—methanol extract.
ᵇDiameter of inhibition zone (mm) including disc diameter of 6 mm.
Antibiotics: S—streptomycin.
Values are the mean of three replicate ($n = 3$).

15.4.1.2 BROTH MICRODILUTION METHOD

Table 15.2 shows the MIC of the tested lichen extracts against the tested bacteria. *R. dumeticola* showed good antibacterial effects toward *B. subtilis, S. aureus,* and *S. epidermidis* with MIC range from 2 to 0.125 mg/mL, while weak toward *E. coli, K. pneumonia,* and *P. aeruginosa* with MIC values ranging from 4 to 16 mg/mL. Extract of *U. baileyi* gives excellent antibacterial effect with lowest MIC value detected toward *B. subtilis* (0.0625 mg/mL) while inhibits other bacteria with MIC values ranging from 0.25 to 16 mg/mL except for *K. pneumonia* that are resistant at all concentration given.

The lichen *H. leucomela* has a weak antibacterial activity of all organisms tested with MIC values ranging from 4 to 16 mg/mL except for *E. coli* that resistant toward acetone extract. The lichens *T. flavicans* manifest moderate antibacterial activity by inhibiting all the tested bacteria within MIC values range from 0.5 to 16 mg/mL except for the *K. pneumonia,* which was resistant to all concentration tested. DMSO had no inhibitory effect on the tested organisms.

TABLE 15.2   Minimum Inhibitory Concentration (MIC) of *Ramalina dumeticola*, *Usnea baileyi*, *Heterodermia leucomela*, and *Teloschistes flavicans* Extracts against the Test Organisms.

Organisms	Minimal Inhibitory Concentration (mg/mL)							
	R. dumeticola		*U. baileyi*		*H. leucomela*		*T. flavicans*	
	A	**M**	**A**	**M**	**A**	**M**	**A**	**M**
Bacillus subtilis	0.25[a]	0.125	0.0625	0.5	4	4	0.5	1
Staphylococcus aureus	1	2	0.5	2	8	16	0.5	1
Staphylococcus epidermidis	1	0.25	0.25	1	8	8	2	8
Escherichia coli	8	4	4	16	-	4	4	16
Klebsiella pneumoniae	16	16	-	-	16	16	-	-
Pseudomonas aeruginosa	16	8	4	16	16	16	16	16

A—acetone extract; M—methanol extract.
[a]Minimum inhibitory concentration (MIC); values given as mg/mL for lichen extract.
Values are the mean of three replicate ($n = 3$).

15.5   DISCUSSION

The antioxidant and antibacterial activities of acetone and methanol extracts of the lichens *R. dumeticola, U. baileyi, H. leucomela, and T. flavicans* were determined. The antioxidant activity has been evaluated based on DPPH (1,1-diphenyl-2-picrylhydrazyl) free radical scavenging and compared with those of commercial antioxidant standards which are ascorbic acid (Vitamin C), BHA, and trolox [(±)-6-hydroxy-2,5,7,8-tetramethylchromane-2-carboxylic acid] [13, 19]. The antioxidant activity of lichen extracts was studied by screening its ability to bleach the stable DPPH radical [2, 15].

From this study, we found that the tested lichen extracts give highest DPPH radical scavenging activity with the increasing amount of total phenolic content in the extracts. Many previous researchers reported that the antioxidant activity of lichen extracts depends on their lipid peroxidation inhibition and total phenol content [3, 17, 18, 20]. The strong correlations between total phenolic contents of tested extracts and the antioxidative activities suggest that phenolic substances might be the major antioxidant compounds in studying extracts as stated by Shahidi and Wanasundara [25]. The presence of the phenolic hydroxyl groups gives phenolic compounds the ability to scavenge radicals [22].

The antibacterial activity of the methanol and acetone extracts of *R. dumeticola, U. baileyi, H. leucomela,* and *T. flavicans* was evaluated by disc-diffusion method and broth microdilution method against *E. coli, P. aeruginosa, K. pneumonia, B. subtilis, S. aureus,* and *S. epidermidis.* The pattern of the antibacterial activity can vary with the species of lichen, its concentration, and the bacteria tested. For both methods, the extract of *R. dumeticola* had the strongest antibacterial activity among the tested species in this study, inhibiting the tested bacteria at low concentrations, while *H. leucomela* showed the lowest activity.

Based on the results, the stronger antibacterial activity possessed by the extract of *R. dumeticola* has a relation with their high phenolic content. This was proved by Adedapo et al. [1], which stated that the presence of different components with antibacterial activity gives consequence to differences in antibacterial activity of different species of lichens, which are phenolic compounds. Usnic acid is the one of the most frequently reported lichen-derived products with a strong antimicrobial activity [9].

The antibacterial pattern of lichen extracts also varies with the bacteria and their cell membrane composition which is different in Gram-positive and Gram-negative bacteria. In this research, most of the tested lichen extracts gave effective antimicrobial activity against Gram-positive bacteria compared with the Gram-negative bacteria due to differences in the composition and permeability of the bacterial cell wall. The cell wall of Gram-positive bacteria is made up of peptidoglycans and teichoic acids, while the cell wall of Gram-negative bacteria is made up of peptidoglycans, lipopolysaccharides, and lipoproteins [7]. These will cause a different mechanism of action of the lichen compounds toward the bacteria.

15.6 CONCLUSION

Based on these results, lichens can be a good and safe, natural antioxidant, and antibacterial agent and could also be of significance in the food industry and pharmaceutical to control various human, animal, and plant diseases. Further studies should be done to isolate and purify the responsible compounds from lichens that exhibit strong antioxidant and antibacterial activities.

15.7 ACKNOWLEDGMENT

This work was funded by Fundamental Research Grant Scheme (FRGS) of Universiti Teknologi MARA (UiTM), Shah Alam. 600-RMI/ST/ FRGS/5/3/Fst(15/2011) "Colourfastness Characteristics of Lichen Coloured Fabrics".

KEYWORDS

- minimal inhibitory concentration
- methanol extract
- acetone extract
- total phenolic content
- DPPH radical scavenging activity
- lichen

REFERENCES

1. Adedapo, A.; Jimoh, F.; Koduru, S.; Afolayan, J. A.; Masika, J. M. Antibacterial and Antioxidant Properties of the Methanol Extracts of the Leaves and Stems of *Calpurina aurea*. *BMC Complem. Altern. Med.* **2008**, *8*, 53–60.
2. Anandjiwala, S.; Bagul, M. S.; Parabia, M.; Rajani, M. 2008. Evaluation of Free Radical Scavenging Activity of an Ayurvedic Formulation, *Panchvalkala. Indian J. Pharm. Sci.* **2008**, *70*, 31–35.
3. Behera, B. C.; Verma, N.; Sonone, A.; Makhija, U. Antioxidant and Antibacterial Activities of Lichen *Usnea ghattensis* in Vitro. *Biotechnol. Lett.* **2005**, *27*, 991–5.
4. Blois, M. S. Antioxidant Determinations by the Use of a Stable Free Radical. *Nature* **1958**, *26*, 1199–1200.
5. Galun, M. *CRC Handbook of Lichenology;* CRC Press: Boca Raton, 1988.
6. Halama, P., Van Haluwin, C. Antifungal Activity of Lichen Extracts and Lichenic Acids. *BioControl* **2004**, *49*, 95–107.
7. Heijenoort, J. Formation of the Glycan Chains in the Synthesis of Bacterial Peptidoglycan. *Glycobiology* **2001**, *11*, 25–36.
8. Hodkinson, B. P.; Lutzoni, F. A Microbiotic Survey of Lichen-associated Bacteria Reveals a New Lineage from the Rhizobiales. *Symbiosis* **2009**, *49*(3), 163–180.
9. Ingolfsdottir, K. Usnic Acid. *Phytochemistry* **2002**, *61*, 729–736.

10. Karagoz, A.; Dogruoz, N.; Zeybek, Z.; Aslan, A. Antibacterial Activity of Some Lichen Extracts. *J. Med. Plants Res.* **2009**, *3*, 1034–1039.

11. Karaman, I.; Sahin, F.; Gulluce, M.; Ogutcu, H.; Sengul, M.; Adiguzel, A. Antimicrobial Activity of Aqueous and Methanol Extracts of *Juniperus oxicedrus*. L. *J Ethnopharmacol.* **2003**, *85*, 231–235.

12. Karthikaidevi, G.; Thirumaran, G.; Manivannan, K.; Anantharaman, P.; Kathiresan, K.; Balasubaramanian, T. Screening of the Antibacterial Properties of Lichen *Roccella belangeriana* (awasthi) from Pichavaram Mangrove (*Rhizophora* sp.). *Adv. Biol. Res.* **2009**, *3*, 127–131.

13. Kekuda, P. T. R.; Vinayaka, K. S.; Praveen Kumar, S. V.; Sudharshanm S. J. Antioxidant and Antibacterial Activity of Lichen Extracts, Honey and Their Combination. *J. Pharm. Res.* **2009**, *2*, 1875–1878.

14. Kirmizigul, S.; Koz, O.; Anil, H.; Icli, S. Isolation and Structure Elucidation of Novel Natural Products from Turkish Lichens. *Turk. J. Chem.* **2003**, *27*, 493–500.

15. Koleva, I. I.; van Beek, T. A.; Linssen, J. P.; Groot, A. D.; Evstatieva, L. N. Screening of Plant Extracts for Antioxidant Activity: a Comparative Study on Three Testing Methods. *Phytochem. Anal.* **2002**, *13*(1), 8–17.

16. Maizura, M.; Aminah, A.; Wan Aida, W. M. Total Phenolic Content and Antioxidant Activity of Kesum (*Polygonum minus*), Ginger (*Zingiber officinale*) and Turmeric (*Curcuma longa*) Extract. *Int. Food Res. J.* **2011**, *18*, 529–534.

17. Manojlovic, N. T.; Vasiljevic, P. J.; Gritsanapan, W.; Supabphol, R.; Manojlovic, I. Phytochemical and Antioxidant Studies of *Laurera benguelensis* Growing in Thailand. *Biol. Res.* **2010**, *43*, 169–176.

18. Odabasoglu, F.; Aslan, A.; Cakir, A.; Suleyman, H.; Karagoz, Y.; Bayir, Y.; Halici, M. Antioxidant Activity, Reducing Power and Total Phenolic Content of Some Lichen Species. *Fitoterapia* **2005**, *76*, 216–219.

19. Paudel, B.; Bhattarai, H. D.; Lee, J. S.; Hong, S. G.; Shin, H. W.; Yim, J. H. Antioxidant Activity of Polar Lichens from King George Island (Antarctica). *Polar Biol.* **2008**, *31*, 605–608.

20. Praveen Kumar, S. V.; Prashith-Kekuda, T. R.; Vinayaka, K. S.; Sudharshan, S. J.; Mallikarjun, N.; Swathi, D. Studies on Antibacterial, Anthelmintic and Antioxidant Activities of a Macrolichen *Parmotrema pseudotinctorum* (des. Abb.). Hale (Parmeliaceae) from Bhadra Wildlife Sanctuary, Karnataka. *Int. J. PharmTech. Res.* **2010**, *2*, 1207–1214.

21. Romagni, J. G.; Dayan, F. E. Structural Diversity of Lichen Metabolites and Their Potential Use. In *Advances in Microbial Toxin Research and Its Biological Exploitation;* Upadhyay R. K, Ed; Kluwar Academic/Plenum Publishers: New York, 2002, pp. 151–169.

22. Sawa, T.; Nakao, M.; Akaike, T.; Ono, K.; Maeda, H. Alkylperoxyl Radical Scavenging Activity of Various Flavonoids and Other Phenolic Compounds: Implications for the Anti-tumor Promoter Effect of Vegetables. *J. Agric. Food. Chem.* **1999**, *47*, 397–492.

23. Selbmann, L.; Zucconi, L.; Ruisi, S.; Grube, M.; Cardinale, M.; Onofri, S. Culturable Bacteria Associated with Antarctic Lichens: Affiliation and Psychrotolerance. *Polar Biol.* **2010**, *33*, 71–83.

24. Singleton, V. L.; Rossi, J. A. Colorimetry of Total Phenolics with Phosphomolybdic-phosphotungstic Acid Reagents. *American J. Enology Vitic.* **1965,** *16*(3), 144–158.

25. Shahidi, F.; Wanasundara, P. K. J. P. D. Phenolic Antioxidants. *Crit. Rev. Food Sci. Nutrit.* **1992,** *32*, 67–103.

26. Stanly, C.; Ali, D. M. H.; Keng, C. L.; Boey, P. L.; Bhatt, A. Comparative Evaluation of Antioxidant Activity and Total Phenolic Content of Selected Lichen Species from Malaysia. *J. Pharm. Res.* **2011,** *4*, 2824–2827.

PART IV
Sustainable Synthesis and Production of Advanced Organomaterials

CHAPTER 16

SYNTHESIS OF UNNATURAL AMINO ACIDS: BETULINIC ACID PEPTIDES AS BIOACTIVE MOLECULES

MOHD TAJUDIN BIN MOHD ALI*, HABSAH BINTI ZAHARI, and SITI AISYAH BINTI ALIASAK

Faculty of Applied Sciences, Universiti Teknologi MARA, Shah Alam, Selangor, Malaysia

Corresponding author. E-mail: tajudinali@salam.uitm.edu.my

CONTENTS

ABSTRACT

Betulinic acid (BA) is one of the triterpene acid compounds showing a remarkable cytotoxicity on various tumor cells. Betulinic acid as the main constituents of *Melaluca cajiput* sp. bark is one of the promising compounds that has anti-retroviral, anti-malaria, anti-HIV, and anti-inflammatory activities. The therapeutic properties of peptides derived from betulinic acid aroused our interest in synthesizing the betulinic acid peptides. We herein report the synthesis of several natural amino acid–BA peptides using standard coupling protocols (HOBt, HBTU). In addition, enantiopure cyclohexene silyl ether–BA peptide was successfully synthesized.

16.1   INTRODUCTION

Over the last 15 years, interest in drugs of plant origin has been reviving and growing steadily, and the drug researchers are exploring the potential of natural products for the cures of still unsurmountable diseases like cancer and AIDS [1]. Betulinic acid (BA) is a pentacylic tripertene natural product initially identified as a melanoma-specific cytotoxic agent that exhibits low toxicity in animal's model [2]. Natural compounds, to treat various types of cancer, have recently attracted considerable interest due to their versatile biological properties and usually broad safety window during administration. One such group of compounds is BA being the best documented example demonstrating cytotoxicity toward several cancer cell lines [3]. BA is a naturally occurring pentacyclic triterpenoid which has antiretroviral, antimalarial, and anti-inflammatory properties, as well as a more recently discovered potential as an anticancer agent, by inhibition of topoisomerase [4]. The structural formula for BA compound is shown in Figure 16.1.

Most of the compounds that can be extracted from the barks of these plants are usually betulin compounds [5]. Betulin compound is similar with BA, but the difference between these two compounds is the functional group at C-28. For BA, the functional group at C-28 is carboxyl group (carboxylic acid), but for betulin the functional group at C-28 is hydroxyl group (alcohol) [6].

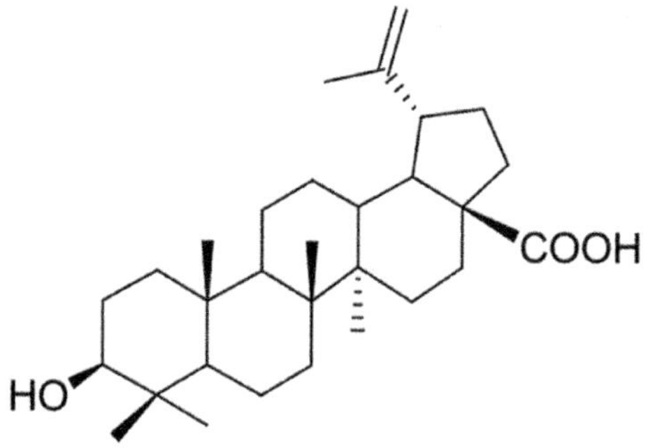

FIGURE 16.1   Structural formula of betulinic acid.

Drag Zalesinski and their coworkers have synthesized several peptide derivatives of BA. They targeted the attachment of amino group such as Boc-Lysine at C-3 BA to afford Lys-BA, **3**. In their synthesis, coupling reaction was carried out using CDI (1,1-carbonyldiimidazole) as depicted in Scheme 16.1.

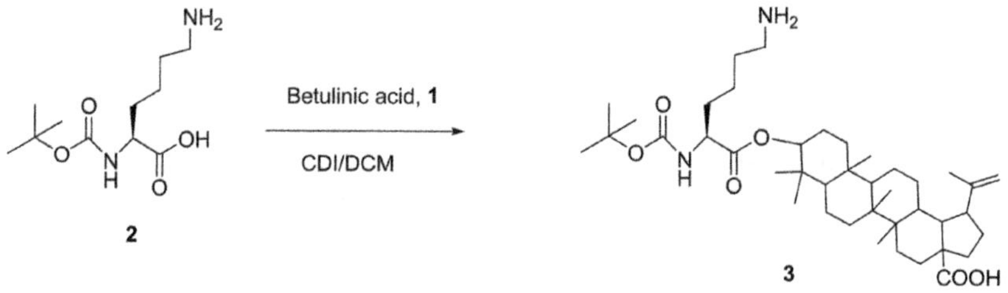

SCHEME 16.1   The synthesis of lycine-betulinic acid.

They found that Lys-BA **3** have, in general, a better cytotoxicity in human pancreatic carcinoma parental and human gastric carcinoma parental cell lines with IC_{50} value 14 and 9.7 μm, respectively[3].

We herein report the synthesis of several natural amino acid–BA peptides using standard coupling protocols (HOBt, HBTU, and DiPEA) as depicted in Scheme 16.2. This work focuses only on the synthesis of BA peptides, which will be the preliminary work in finding novel bioactive molecules.

16.2   MATERIALS AND METHODS

16.2.1   GENERAL EXPERIMENTS FOR PEPTIDE SYNTHESIS

S solution of BA (0.016 g, 0.035 mmoles) in DCM (2 mL) HOBt (0.00713 g, 1 equiv.), followed by HBTU (0.02 g, 1.5 equiv.) and natural amino amine or amine salt (Phe-Ala, Leu-Ala, cyclohexene silyl ether) (0.02 g, 1 equiv.) at 0°C. Then, DIPEA was added until the solution turned into basic. Then, the ice bath was removed and the reaction mixture was stirred for 24 h at room temperature. The solvent was removed by rotary evaporator. The white solid then was dissolved in 5 mL EtOAc and was filtered. The filtrate was washed by 1 M HCl (2 mL), saturated NaHCO$_3$ (5 mL), and brine (5 mL). The organic layer was dried over by MgSO$_4$ and the solvent was evaporated under reduced pressure to give white solid crude product, which was purified by column chromatography on silica gel (petroleum: ethyl acetate, 4:6).

16.3   RESULTS AND DISCUSSION

The synthesis of several natural amino acids with BA using standard coupling protocol involved several coupling reagents, that is, HOBt, HBTU, and a strong base, DIPEA. The reaction was carried out at 0°C for 24 h to obtain Alanine–BA **2a** (Ala-BA), Penylalanine–BA **2b** (Phe-ala-BA), Leu-Ala-betulinic **2c** (Leu-Ala-BA) peptides compounds with moderate yields.

Reagent and conditions: BA (1 equiv.), EtOH, TFA, DCM, 0°C, DCM, HOBt (1.1 equiv.), HBTU (1 equiv.), amino acids or cyclohexene silyl ether (1.1 equiv.), DiPEA (1.1 equiv. 0°C), 24 h, % yields (2a: 85%, 2b: 86%, 2c: 70%, and 2d: 69%).

In addition, enantiopure cyclohexene silyl ether-BA peptide was successfully synthesized. Cyclohexene silyl ether was envisioned from racemix epoxide 3, which had undergone epoxide ring opening using salen complex catalyst to give azido cyclohexene. By using reduction amination and deprotection protocols, enantiopure cyclohexene was obtained in a moderate yield (68%).

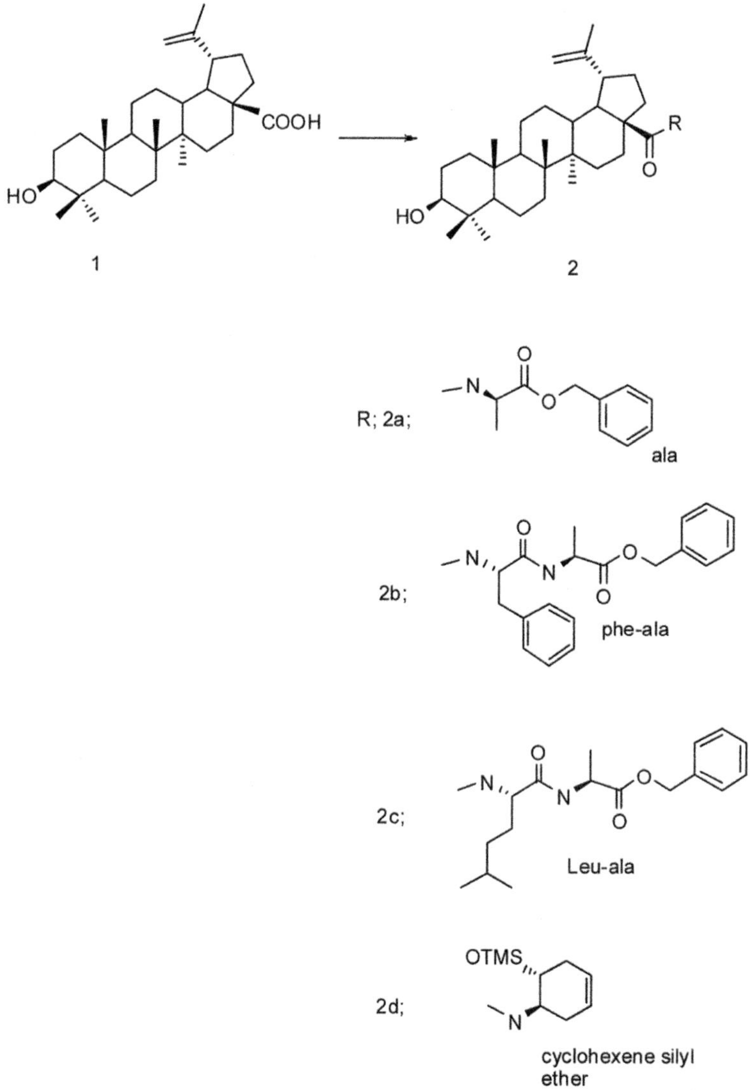

SCHEME 16.2 The synthesis of peptide-betulinic acid.

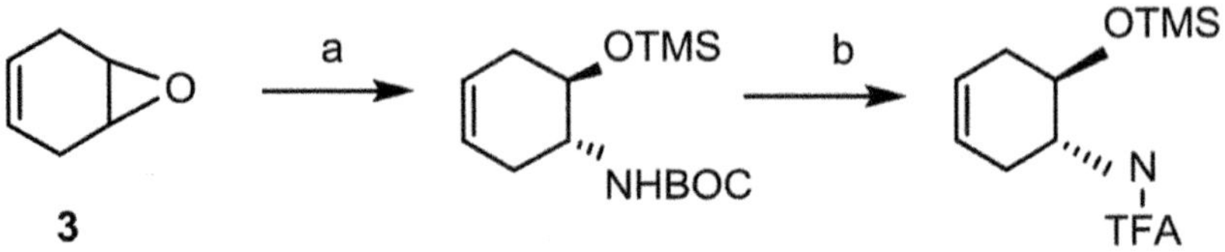

SCHEME 16.3   Synthesis of enantiopure cyclohexene silyl ether.

Reagent and conditions: a. (*R,R*) Salen complex 2 (2 mol%), TMSN$_3$ (1.05 equiv.), Et$_2$O, rt, 46 hrs, 68%, 85% ee. b. TFA in DCM (30%), 0°C, 3 h, quant.

16.3.1 SPECTROSCOPIC DATA FOR SYNTHESIZED COMPOUNDS

(2a) (2*R*)-benzyl 2-1*R*,3a*S*,5a*R*,5b*R*,9*S*,11a*R*)-9-hydroxy-5a,5b,8,8,11a-pentamethyl-1-(prop-1-en-2-yl)icosahydro-1H- cyclopenta[a]chrysene-3a-carboxamido)propanoate

R*f* = 0.3 (SiO$_2$, hexane/ethylacetate 4:1); Yields: 85%

¹H-NMR (300 MHz CDCL$_3$); δ = 8.11-8.21 (d,1H,NH), 7.46-7.23 (m, 5H, Bn), 5,21-5.11 (d,2H, *J* = 4 Hz, CH$_2$Bn), 4.73(s,1H),4.61(s,1H), 4.55-4.42 (m, 1H, CHN), 3.28-3.11 (m, 1H, CH-OH), 3.10- 2.99 (m,1H,CH), 1.56-1.40(m,3H, CH$_3$-ala), 1.81(s,3H, CH$_3$), 1,21 (s,3H, CH$_3$), 0.98 (s,3H, CH$_3$), 0.96 (s,3H, CH$_3$), 0.81 (s,3H, CH$_3$), 0.76 (s,3H, CH$_3$),

¹³C-NMR (75.5 MHz, CDCl$_3$): **¹³C-NMR** (75.5 MHz, CDCl$_3$): δ = 175.8, 172.0, 150.9 (C=C, BA), 136.4, 128.7-128.5 (Ph), 109.5 (C=C, BA), 78.9 (C–OH), 65.7 (CH$_2$Ph), 56.2 (CHN-ala), 46.1 (C–H), 28.2 (CH$_3$-ala), 28.0, 19.4, 16.5, 15.8, 15.4, 14.6, MS [Cl, NH$_3$]: m/z (%) 618.1 [M + H⁺]

(2b) (2*S*)-benzyl2-((2*S*)-2-((1*R*,3a*S*,5a*R*,5b*R*,9*S*,11a*R*)-9-hydroxy-5a,5b,8,8,11a-pentamethyl-1-(prop-1-en-2-yl)icosahydro-1H-cyclopenta[a]chrysene-3a-carboxamido)-3-phenylpropanamido)propanoate

R*f* = 0.26 (SiO$_2$, hexane/ethylacetate 4:1); Yields: 86%

¹H-NMR (300 MHz CDCL₃); δ = 8.11-8.21(s,1H,NH), 7.30-7.61 (m,10H, Ph), 5.26-5.18 (d, 2H, J = 4.3 Hz, CH₂Bn), 4.73 (s,1H), 4.61(s,1H), 4.66-4.83 (m, 2H, CHNH, CHNH), 2.75-2.60 (m,1H,CH₂CH), 2.50-2.43 (m,1H,CH₂CH), 3.28-3.11 (m, 1H, CH-OH), 3.10-2.99 (m,1H,CH), 1.81(s,3H, CH₃), 1.56-1.40(m,3H, CH₃-ala),1,21 (s,3H, CH₃), 0.98 (s,3H, CH₃), 0.96 (s,3H, CH₃), 0.81 (s,3H, CH₃), 0.76 (s,3H, CH₃)

¹³C-NMR (75.5 MHz, CDCl₃): δ = 172.0, 156.2, 150.9 (C=C, BA), 136.4, 128.7-128.5 (Ph), 109.5 (C=C, BA), 78.9 (C-OH), 67.6 (CHNH-Ph), 65.7 (CH₂Ph), 56.2 (CHN-ala), 46.1 (C–H), 28.2, 28.0, 19.3, 16.5, 15.6, 15.7, 14.5. MS [Cl, NH₃]: m/z (%) 769.8 [M + NH₄⁺].

(2c) (2S)-benzyl 2-((2S)-2-((1R,3aS,5aR,5bR,9S,11aR)-9-hydroxy-5a,5b, 8,8,11a-pentamethyl-1-(prop-1-en-2-yl)icosahydro-1H-cyclopenta[a] chrysene-3a-carboxamido)-5-methylhexanamido)propanoate

Rf = 0.74 (SiO₂, hexane/ethylacetate 4:1); Yields: 70%

¹H-NMR (300 MHz CDCL₃); δ = 7.42-7.31(m,5H,Bn), 5.26-5.18 (d, 2H, J = 4.6 Hz, CH₂Bn), 4.73(s,1H), 4.61(s,1H), 4.66-4.83 (m, 2H, CHNH, CHNH), 3.28-3.11 (m, 1H, CH-OH), 3.10- 2.99 (m,1H,CH), 2.25-1.83 (m,2H,CH₂CH), 1.81(s,3H, CH₃), 1.56-1.40(m,3H, CH₃-ala),1,21 (s,3H, CH₃), 0.98 (s,3H, CH₃), 0.96 (s,3H, CH₃), 0.81 (s,3H, CH₃), 0.76 (s,3H, CH₃)

¹³C-NMR (75.5 MHz, CDCl₃): δ = 172.0, 156.2, 150.9 (C=C, BA), 136.4, 128.7- 128.5 (Ph), 109.5 (C=C, BA), 78.9 (C–OH), 67.7 (CHNH), 65.7 (CH₂Ph), 56.2 (CHN-ala), 46.1 (C–H), 28.2, 28.0, 19.3, 16.5, 15.6, 15.7, 15.2 (CH₃), 14.5. MS [Cl, NH₃]: m/z (%), 745.7 [M + H⁺]

(2d) (1R,3aS,5aR,5bR,9S,11aR)-9-hydroxy-5a,5b,8,8,11a-pentamethyl-1-(prop-1-en-2-yl)-N-((1R,6R)-6-trimethylsilyloxy)cyclohex-3-enyl)icosa-hydro-1H-cyclopenta[a]chrysene-3a-carboxamide

Rf = 0.66 (SiO₂, hexane/ethylacetate 4:1); Yields: 69%

¹H-NMR (300 MHz CDCL₃); δ = 7.30-7.21(m,5H,Ph), 5.35-5.23(m, 2H, CH-Olefin), 4.73(s,1H),4.61(s,1H), 3.66-3.75 (m, 1H, CHO), 3.53-3.65 (m, 1H, CHN), 2.26-2.55 (m, 2H, CH₂), 1.69 (s,3H), 0.97 (s,3H), 0.96 (s,3H), 0.93 (s,3H), 0.82 (s,3H), 0.75 (s,3H), 0.0 (s,9H,OTMS)

^{13}**C-NMR** (75.5 MHz, CDCl$_3$): δ = 172.0, 150.9 (C=C, BA), 136.4, 124.1 (C=C), 109.5 (C=C, BA), 78.8, 78.9 (C–OH), 50.6 (CH) 46.1 (C–H), 32.9 (CH$_2$-Olefin), 29.9 (CH$_2$-Olefin), 28.0, 19.4, 16.5, 15.8, 15.4, 14.6MS [Cl, NH$_3$]: m/z (%) 624.1 [M + H$^+$]

16.4   CONCLUSION

The peptides of amino acids–BA were successfully synthesized using standard peptides coupling protocol. All synthesized molecules will be subjected to bioassay evaluations such as antituberculosis activity, anti-cancer activity, and antibacterial activity in the near future.

16.5   ACKNOWLEDGMENT

The authors thank UiTM, Shah Alam, for an UPTA grant.

KEYWORDS

- synthesis
- peptides
- betulinic acid
- amino acids
- bioactive molecules

REFERENCES

1. Yogeeswari, P.; Sriram, D. Betulinic Acid and its Derivatives: A Review on their Biological. *Curr. Med. Chem.* **2005**, *12*, 657–666.
2. Sudhakar, C.; Sabitha, P.; Shashi, K. R.; Stephen, S. Betulinic Acid Inhibits Prostate Cancer Growth Through Inhibition of Specificity Protein Transcription Factors. *Cancer Res.* **2007**, *67*, 2816–2823.
3. Malgorzata, D. Z; Julita, K.; Jolanta, S.; Teresa, W.; Maciej, Z.; Pawel, S.; Marcin, D. Ester of Betulin and Betulinic Acid with Amino Acids Have Improved Water

Solubility and Are Selectively Cytotoxicity Toward Cancer Cells. *Bioorg. Med. Chem. Lett.* **2009,** *519,* 4814–4817.

4. Shun-ichi, W.; Reiko, T. Betulinic Acid and Its Derivatives, Potent DNA Topoisomerase II Inhibitors, from the Bark of *Bischofia javanica. Chem. Biodiv.* **2005,** *2*(5), 689–694.

5. Abdul M. Md.; Syed, M. T.; Apurba, S. A.; Debasish, B.; Mohamad, S. I. Isolation and Identification of Compound from the Leaf of Extract of *Dillenia indica* Linn. *Bangladesh Pharm. J.* **2010,** *1*(13), 49–53.

6. Ali, M.; Tajudin, M.; Jasmani, H.; Yasin, Y. *Chemical Modification of Betulinic Acid*; CSSR Research, Institute of Research, Development and Commercialization (IRDC), UiTM, Shah Alam, 2012.

CHAPTER 17

SYNTHESIS OF ENANTIOPURE AZIDO TRIMETHYLSILOXY CYCLOHEXENE DERIVATIVES: A USEFUL INTERMEDIATES FOR THE SYNTHESIS OF TAMIFLU

MOHD TAJUDIN MOHD ALI*, SITI AISYAH ALIASAK, HABSAH ZAHARI, and SYED ABDUL ILLAH ALYAHYA SYED ABDUL KADIR

Faculty of Applied Sciences, Universiti Teknologi MARA, Shah Alam, Selangor, Malaysia

*Corresponding author. E-mail: tajudinali@salam.uitm.edu.my

CONTENTS

ABSTRACT

Peptides bearing natural amino acids scaffolds are widespread in nature and known to exhibit a wide array of biological activities. As part of research endeavor in investigating the potential of enantiopure azido cyclohexene silyl ether, we hereby present an approach toward the intermediate synthesis of anti-influenza drug, Tamiflu, starting from azido cyclohexene silyl ether. The methodology involved reduction amination of azide, followed by Boc protection of amine termini. Deprotection of Boc group and coupling reaction with natural amino acids using standard peptide coupling procedure afforded enentiopure cyclohexene peptides.

17.1 INTRODUCTION

Synthetic peptides are widely used either in medical research or in the diagnosis of diseases as they are biologically active. Many synthetic peptides have been significantly established in accord with the advanced peptide coupling reactions. Evolution of the new peptide compounds along with diverse biological activities will significantly enhance the research pool of known biologically active compounds [1, 2].

Every year, influenza viruses cause global epidemics that result in significant mortality and morbidity. Previous studies stated that Oseltamivir phosphate **1** (Tamiflu) is an approved orally active neuraminidase inhibitor used for treatment of human influenza and H5N1 avian flu [3, 4]. This drug has pediatric indications approved by the Food and Drug Administration [5]. Oseltamivir phosphate **1** an ester-type pro-drug is metabolize to its active form Oseltamivir carboxylate **2** by ester hydrolysis reaction, after orally administration. Oseltamivir phosphate **1** inhibits influenza virus infection and replication in vitro, while Oseltamivir carboxylate **2** inhibits influenza A and B [6]. The natural product precursor in synthesis of oseltamivir is *Illicium verum*, which is consider as a primary source of shikimic acid (precursor to oseltamivir). Additional natural product sources of shikimic acid include *Liquidambar* spp. [7].

Previous studies stated that the synthesis of oseltamivir depends on the supply of raw materials [8,9]. Therefore, it is crucial to develop various synthetic methodologies and intermediates toward the synthesis of oseltamivir. Several novel oseltamivir derivatives (OS-11 **3**, OS-20 **4**, and OS-23 **5**) were successfully synthesized and exhibit antiviral activity against

H1N1 and H3N2 virus. Based on the cytotoxicity studies, compounds OS-11 **3**, OS-20 **4**, and OS-23 **5** had similar toxicity with oseltamivir phosphate **1** and oseltamivir carboxylate **2** with IC_{50} value of 0.562 mg/ml, 0.501 mg/ml, and 0.562 mg/ml, respectively [10].

In this studies, synthetic of enantiopure azido trimethylsiloxy cyclohexene **5** with selected amino acids as an intermediate compounds toward the synthesis of Oseltamivir have been synthesized in efficient manner using standard peptides coupling protocol.

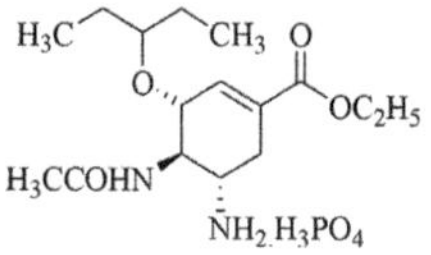

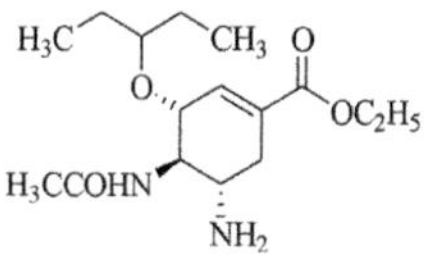

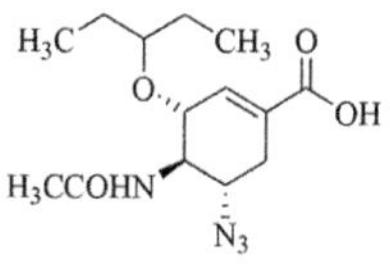

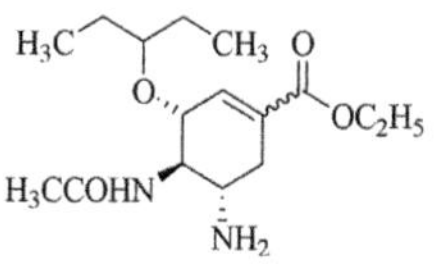

17.2 MATERIALS AND METHODS

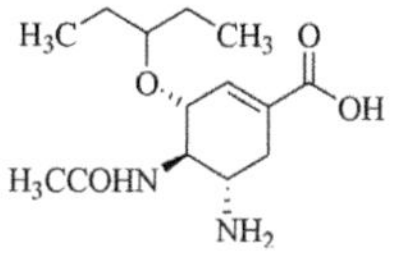

Synthesis pathways to azido trimethylsiloxy cyclohexene

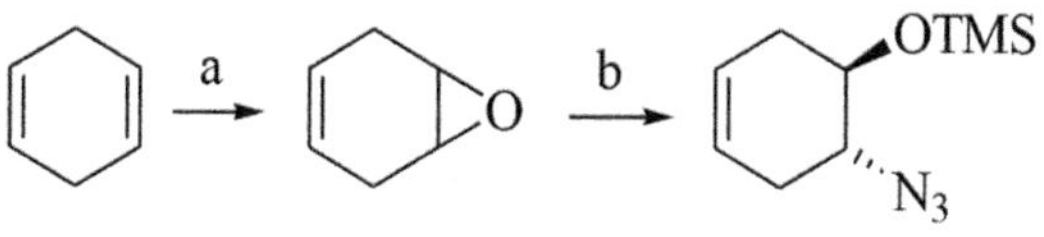

Reagents and conditions (a) *m*-CPBA (1 equiv.), K_2HPO_4 (1 equiv), DCM, 89% (b) $TMSN_3$, Salen Cr(III) complex, Et_2O, rt, 46 hrs, 68%, 86% ee

SCHEME 17.1 The synthesis of azido trimethylsiloxy cyclohexene **(8)**.

Synthesis pathways to azido trimethylsiloxy cyclohexene peptides

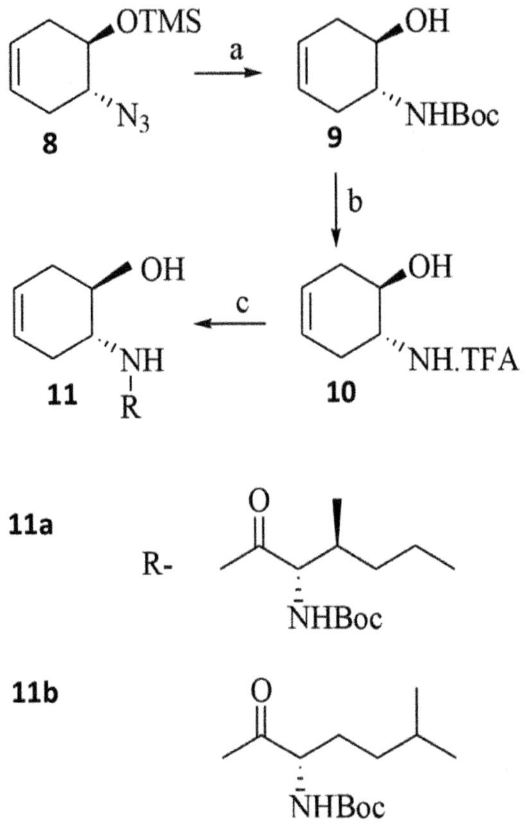

Reagents and conditions (a) Boc$_2$O (1.5 equiv.), Pd(OH)$_2$/C, Triethylsilane (1.1 equiv.), EtOH, 46% (b) TFA, DCM, 0°C, 3 h (c) i. THF, HOBt (1.1 equiv.), HBTu (1 equiv.) DiPEA (1.1 equiv., 0°C), Boc-D-Ile-OH (1.1 equiv.), 57% or Boc-D-Leu-OH (1.1 equiv) 64%.

SCHEME 17.2   The synthesis of azido trimethylsiloxy cyclohexene peptides.

(2-Hydroxy-cyclohexyl)-carbamic acid tert-butyl ester (9)

To a stirred mixture of azido trimethylsiloxy cyclohexene **8** (0.2 g, 1.02 mmol, 1 equiv.) in 6 mL ethanol, was added *tert*-butoxycarbonyl (Boc$_2$O) (0.33 g, 1.5 mmol, 1.5 equiv.) and 20% Pd(OH)$_2$/C (10.4 mg) at room temperature. Then triethylsilane (0.28 mL, 1.74 mmol, 1.7 equiv.) was added sequentially, and the mixture was stirred for 20 h under N$_2$

atmosphere. The mixture was filtered through Celite, and the filtrate was concentrated under reduce pressure to remove the solvent to give yellowish solid which was purified by column chromatography on silica gel (petroleum ether:ethyl acetate, 15:1) to yield 0.126 g, 46% of compound **9** as white solid.

White solid. R_f = 0.275 (SiO$_2$, hexane/ethylacetate 15:1); **^{1}H-NMR** (300 MHz, CDCl$_3$): δ = 5.59 (d, 2H, C*H* cyclohexene), 4.67 (s, 1H, NH), 3.70 (s, 1H, C*H*NHBoc), 3.32 (s, 1H, C*H*O), 2.49–2.56 (d, 2H, C*H*$_2$ cyclohexene), 1.46–1.49 (m, 9H, Boc), 0.78 (s, 1H, OH); **^{13}C-NMR** (75.5 MHz, CDCl$_3$): δ = 159.16 (CO, Boc), 124.28 (C=C), 76.62 (C-Boc), 65.09 (CNHBoc), 30.9 (C*H*$_2$CHNBoc), 28.37 (C*H*$_2$CHOH), 24.71 (Boc-C). MS [Cl, NH$_3$]: m/z (%) 214.3 [M + NH$_4$$^+$]

2,2,2-Trifluoro-N-(6-hydroxycyclohex-3-enyl)acetamide (10)

A stirred compound **9** (31 mg, 0.15 mmol) was treated with saturated HCl in dry ethylacetate (8 ml) at 0°C for 3 h. Then the solvent was removed in vacuo, dried on oil pump to afford **10**, which was used further for the next reaction.

[1-(6-Hydroxy-cyclohex-3-enylcarbamoyl)-2-methyl-butyl]-carbamic acid tert-butyl ester (11a)

A solution of azido carboxylic acid **10** (30.0 mg, 0.16 mmol) was stirred in dry THF (2 mL). HOBt (11.2 mg, 1.1 equiv.), HBTU (30.0 mg, 1.0 equiv.), and Boc-*D*-Ile-OH (20.0 mg, 1.1 equiv.) were added in dry THF at 0°C. The ice bath was removed and the reaction mixture was stirred for 24 h at room temperature. The solvent was removed by rotary evaporator. Then, the reaction mixture was dissolved in EtOAc (5 mL) and was washed with 1 M HCl (5 mL), 5% NaHCO$_3$ (5 mL), and brine (5 mL). The organic layer was dried over by Na$_2$SO$_4$ and the solvent was evaporated under reduced pressure. The crude product was purified by column chromatography on silica gel, yielding 27.0 mg, 57% of expected product as white solid.

White solid. R_f = 0.375 (SiO$_2$, hexane/ethylacetate 15:1); **^{1}H-NMR** (300 MHz, CDCl$_3$): δ = 5.6 (d, 2H, C*H* cyclohexene), 6.29 (s, 1H, N*H*Boc), 5.18 (s, 1H, CN*H*), 4.04 (s, 1H, C*H*OH), 3.83 (s, 1H, C*H*N), 2.5–2.55 (d,

1H, C*H*CH$_3$CH$_2$C(NHBoc)), 1.74–2.06 (m, 4H, CH$_2$), 1.45–1.47 (m, 9H, Boc), 1.29 (s, 2H, CH$_2$), 0.85–0.97 (m, 6H, CH$_3$); ^{13}**C-NMR** (75.5 MHz, CDCl$_3$): δ = 159.16 (CO, Boc), 124.1 (C=C), 80.32 (C-Boc), 81.59 (COH), 74.4 (CNH), 70.08 (CNHBoc), 36.77 (CH), 28.32 (Boc-C), 24.89 (CH$_2$), 15.53 (CH$_3$), 11.81 (CH$_3$). MS [Cl, NH$_3$]: m/z (%) 344.5 [M + H$^+$]

4-Amino-4-(6-hydroxy-cyclohex-3-enylcarbamoyl)-2-methyl-butyric acid tert-butyl ester (11b)

Solution of azido carboxylic acid **10** (30.0 mg, 0.16 mmol) was stirred in dry THF (2 mL). HOBt (11.2 mg, 1.1 equiv.), HBTU (30.0 mg, 1.0 equiv.), and Boc-*D*-Leu-OH (20.0 mg, 1.1 equiv.) were added in dry THF at 0°C. The ice bath was removed and the reaction mixture was stirred for 24 h at room temperature. The solvent was removed by rotary evaporator. Then, the reaction mixture was dissolved in EtOAc (5 mL) and was washed with 1 M HCl (5 mL), 5% NaHCO$_3$ (5 mL), and brine (5 mL). The organic layer was dried over by Na$_2$SO$_4$ and the solvent was evaporated under reduced pressure. The crude product was purified by column chromatography on silica gel, yielding 30.0 mg, 63% of expected product as white solid.

White solid. R$_f$ = 0.375 (SiO$_2$, hexane/ethylacetate 15:1); ^{1}**H-NMR** (300 MHz, CDCl$_3$): δ = 5.0 (d, 2H, C*H* cyclohexene), 4.25–4.38 (m, 1H, C*H*NHBoc), 3.33–3.48 (m, 2H, CH cyclohexene), 1.9–2.1 (d, 2H, C*H*$_2$CHOH), 1.67–1.76 (d, 2H, C*H*$_2$CHN), 1.48 (s, 9H, Boc), 0.9–1.0 (m, 6H, CH$_3$); ^{13}**C-NMR** (75.5 MHz, CDCl$_3$): δ = 156.0 (CO, Boc), 124.5 (C=C), 81.57 (COH), 33.6 (CH), 28.32 (Boc-C), 24.81 (CH$_2$), 22.88 (CH$_3$). MS [Cl, NH$_3$]: m/z (%) 341.3 [M + H$^+$]

17.3 RESULTS AND DISCUSSION

In this article, we wish to report an efficient approach toward the synthesis of peptides starting from azido cyclohexene silyl ether which was previously reported by M. Ali *et al.* [2]. The synthesis starts from commercially available 1,4-cyclohexadiene **6**.

Epoxidation of 1,4-cyclohexadiene **6** is performed in DCM using m-CPBA as a strong oxidizing agent in the presence of K$_2$HPO$_4$ yielded

1,4-cyclohexadiene monoxide **7**. The reaction was continued by enantioselective assymetric ring opening of epoxide catalyzed by Salen–Cr(III) complex to furnish azido trimethylsiloxy cyclohexene **8** as the starting material. Reduction of azide group furnished *N*-Boc protected amine termini compound which undergoes deprotection of Boc group and coupling reaction with selected natural amino acids [1].

The synthesis of target compound **11a**, **11b** began with azido trimethylsiloxy cyclohexene which was derived from ring opening of cyclohexene epoxide with $TMSN_3$.[1b] To afford **9**, azido functionality was transformed to NHBoc by reduction amination using $Pd(OH)_2$ in charcoal, Boc anhydride, and triethylsilane as a proton donor. Deprotection of Boc group in compound **9** by treating with TFA in DCM at 0°C gave cyclohexene trifluoro acetic acid salt **10**. Coupling reaction of compound **10** with selected natural amino acids (*N*Boc-isoleucine or *N*Boc-leucine) furnished two cyclohexene peptides **11a**, **11b**.

As a result, the optimization of the envisaged structures of selected dipeptide derivatives is expected to increase the understanding biological activity relationship in this study. Several steps needed to reach the target molecule Oseltamivir, which involves Mitsunobu reaction and protection of amine, allylic oxidation reaction by selenium oxide, protection of hydroxyl group, Michael addition reaction with cyano group, and hydrolysis reaction.

17.4 CONCLUSION

We have successfully synthesized two derivatives of azido trimethylsiloxy cyclohexene with moderate yields using peptide coupling protocols. Further chemical transformations of the synthesized compounds are needed to reach the target compound (Tamiflu).

17.5 ACKNOWLEDGMENT

The authors are thankful to the Universiti Teknologi MARA Malaysia (grant no: 600-RMI/DANA 5/3/RIF 690/2012 and FRGS grant no: 600-RMI/FRGS 5/3(3/2013) for financial support.

Bioresources Technology in Sustainable Agriculture

KEYWORDS

- Tamiflu
- enantiopure azido trimethylsiloxy cyclohexene
- oseltamivir
- peptide synthesis
- influenza

REFERENCES

1. Ali, M. T. M. Synthesis of (-)-Geissman Waiss Lactone, cis γ-Butyrolactone Derivatives and γ-Peptide. *Worldcat*. **2012**, 80–113.
2. Ali, M. T. M; Macabeo, A. P. G.; Wan, B.; Franzblau, S.; Reiser, O. Asymmetric Approach Toward New Conformationally Constrained Cis Gamma Butyrolactone. In Abstract Of Papers, 245th ACS National Meeting & Exposition: New Orleans, 2013.
3. Raghavan, S.; Babu, V. S. Enantioselective Synthesis of Oseltamivir Phosphate. *Tetrahedron*. **2011**, *67*(11), 2044–2050.
4. Wang, S. Q.; Mita, T.; Cheng, X. C.; Dong, W. L.; Wang, R. L.; Chou, K. C. Three New Powerful Oseltamivir Derivatives for Inhibiting the Neuraminidase of Influenza Virus. *Biochem. Biophys. Res. Commun.* **2010**, *401*, 188–191.
5. Kim, H.; Park, K. A New Efficient Synthesis of Oseltamivir Phosphate (Tamiflu) from (−)– Shikimic Acid. *Tetrahedron Lett.* **2012**, *53*(13), 1651–1563.
6. Vishkaee, T. S.; Mohajerani, N.; Nafisi, S. A Comparative Study of the Interaction of Tamiflu and Oseltamivir Carboxylatewith Bovine Serum Albumin. *J Photochem Photobiol*. **2013**, *119*, 65–70.
7. Avula, B.; Wang, Y. H.; Smillie, T. J.; Khan, I. J. Determination of Shikimic Acid in Fruits of Illicium Specieces and Various Other Plant Samples. *Chromatographia*. **2009**, *69*, 307–314.
8. Satoh, N.; Akiba, T.; Yokoshima, S.; Fukuyama, T. A. Practical Synthesis of (−)– Oseltamivir. *Tetrahedron*. **2009**, *65*, 3239–3245.
9. Nie, L. D.; Shi, X. X. A Novel Asymmetricsynthesis of Oseltamivir Phosphate (Tamiflu) from (−)– Shikimic Acid. *Tetrahedron : Asymmetry*. **2009**, *20*, 124–129.
10. Janusz, K.; Marcin, K.; Justyna, J.; Magdalena, K.; Michal, B. Antiviral Activity of Novel Oseltamivir Against Some Influenza Virus Strain. *Acta. Biochimica. Polonica*. **2014**, *61*(3), 509–513.

CHAPTER 18

SYNTHESIS OF AMINO-BASED METAL ORGANIC FRAMEWORK (MOF) IN OXIDATIVE CATALYSIS

KARIMAH KASSIM[1,*], SITI NURHAZLIN JALUDDIN[2], and WAN NAZIHAH WAN IBRAHIM[2]

[1]*Institute of Science, Universiti Teknologi MARA, 40450 Shah Alam, Selangor, Malaysia*

[2]*Faculty of Applied Sciences, Universiti Teknologi MARA, 40450 Shah Alam, Selangor, Malaysia*

Corresponding author. E-mail: karimah@salam.uitm.edu.my

CONTENTS

ABSTRACT

Amino-based metal organic frameworks (MOFs) derived from dimethyl 5-aminoisophthalate with Mn(II) acetate were successfully synthesized using synthetic thermolysis method. The synthesis ligands of MOFs were characterized by Fourier transform infrared (FTIR) and scanning electron microscopy (SEM). Mn-MOF was tested as oxidative catalyst for 3-nitrobenzaldehyde and 2,4-hydroxybenzaldehyde.

18.1 INTRODUCTION

Metal-organic frameworks (MOFs) are porous solids constructed from metal ions and multidentate organic ligands. It consists of two major components which are organic ligands (struts) and the metal centers (joints). Both components are connected by chemical bonds and intermolecular interactions, resulting in networks topologies with well defined crystalline structure. MOFs have attracted great interest in catalysis and industrial application due to their potential application as versatile and topology structures [1]. However, the use of MOFs in heterogeneous catalysis is one of the most challenging fields [2]. Therefore, more research in this field is needed.

Over the years, the use of catalysts in the manufacturing industry had increased dramatically due to the industries' demand. One of the issues is the cost of recyclability and the difficulty in recovery. Recovery and reusability of the catalyst may enhance the economy of the processes and also decrease contamination of products with residual metal species. Therefore, the introduction of MOFs as active and selective organocatalyst will overcome this situation. Although much research recently has been devoted to designing MOFs structures, the challenge to produce target network structures from the simple reaction of metal ions and organic linkers are still enormous.

In catalysis and gas storage fields, it is necessary to design the MOFs with desired topology before adding the functional group to the framework. A numbers of applications are required to add functional groups to modify surface property and pore geometry [3]. MOFs can be tailored to become active and selective solid organocatalyst via functionalize with organic linkers. Previous studies [4] reported that most functionalities are Lewis and Bronsted basic moieties of amine, amides or urea derivatives groups.

According to Despande (2012)[5], amino functional groups had hindered MOF formation when left exposed during crystal growth.

Therefore, protect ion is needed. A di-*tert*-butyl dicarbonate (Boc) group serves to mask an amino functional group that is found to otherwise preclude MOF growth as well as to avoid steric clashes between NHBoc groups of neighboring ligands. Recently, catalytic oxidation has become an important technology that finds application in chemical industries [6]. Ideally, the pore size of MOFs plays an important role in catalytic reactions. Ma and Meng (2010) [7] reported that porous MOFs can serve as ideal host or catalyst support. Thus, further research in this field is important.

Apart from that, the modular nature of MOFs synthesis also makes it possible to introduce two (or more) different catalytic sites into a single MOFs material. For example, one active site can be incorporated into the MOFs as part of the bridging ligand, whereas the other active site can be introduced in the metal-connecting points. A series of isoreticular MOFs built from chiral Mn salen-derived dicarboxylate-bridging ligands performs succesfully in sequential asymmetric epoxidataion (via manganese centres) and epoxide ring-opening reactions (via zinc centres) [8]. The control experiments confrims that two active sites do not interfere with each other, owing to the isolation of the active sites within the framework. With further study, practically useful multiple active sites of heterogeneous catalysts could arise from this molecular building block approach.

This study emphesizes on post-synthetic thermolysis method. Ideally, the catalytic functions can be introduced into hybrid MOFs via post-synthetic thermolysis, which is formation of linker will occur during the synthesis of MOFs. This method will mask an amino functional group that hindered MOFs formation when left expose during the crystal growth steps. Besides that, this study reports the synthesis of amino-based MOFs for the catalytic application of manganese(II) acetate. The synthesis of amino-based MOF using the Boc for protection of amino group and complexation of the metal linker in the framework is mentioned in the methodology section. Characterization of MOF and the results for the catalytic reaction was discussed in findings sections.

18.2 EXPERIMENTAL SECTIONS

18.2.1 MATERIALS AND REAGENTS

Dimethyl-5-aminoisophthalate (Sigma Aldrich, D5P), manganese (III) acetate dihydrate (Sigma Aldrich, 98%, $Mn(Aoc_3)_2.2H_2O$, triethylamine

(Sigma Aldrich, 99%, Et₃N)(TEA), di-*tert*-butyl dicarbonate (Sigma Aldrich, Boc), 4-dimethylaminopyridine (Merck, 99% DMAP), N,N dimethylformamide (Merck, 99.8% DMF), potassium hydroxide (KOH), lithium bromide(LiBr), hydrochloricacid (Merck, HCl), tetrahydrofluoron(Merck, THF), acetonitrile, dichloromethane, 3-nitrobenzaldehyde and 2,4-hydroxybenzaldehyde.

18.2.2 METHODOLOGY

18.2.2.1 SYNTHESIS OF METAL LINKER LIGAND

In this step, the metal linker ligand was synthesized by addition of Boc. The reason for using Boc is to protect the amino group of the starting materials. Figure 18.1 shows the synthesis of MOFs linker ligand. The method was performed followed the reported procedures [5].

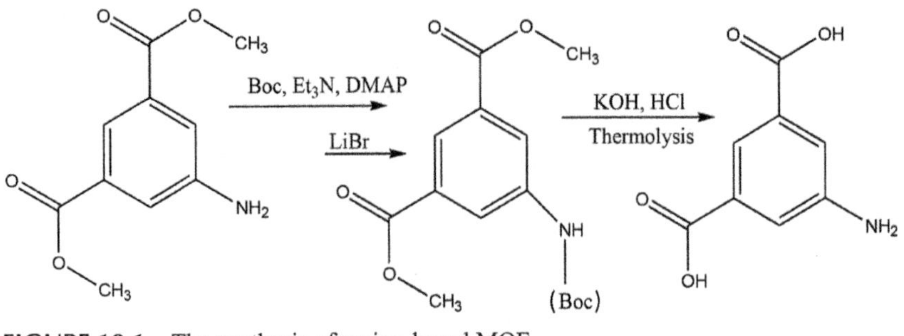

FIGURE 18.1   The synthesis of amino-based MOF.

Dimethyl 5-aminoisophthalate (272 mg, 0.1 mol), TEA (6 mL), DMAP (0.6 g) and 4.5 g Boc were added into THF (40 mL) and stirred at 40 °C for 6 hours followed by removal of excess solvent by rotary evaporation. The conversion to dimethyl 5-[bis(*tert*-butoxycarbonyl)amino] benzene-1,3-dicarboxylate occurred at this stage. The reaction mixture was then diluted with acetonitrile (20 mL) and LiBr (4 mL) and stirred vigorously for 9 hours at 65°C. All volatiles were removed and the crude product in dichloromethane was filtered. The solid formed was washed with hexane and dried.

A solution of the product was added in THF (6 mL) and 1 M KOH (2 mL) followed by reflux for 12 hours. After cooling to room temperature, the THF was removed by rotary evaporation and the solution was cooled in ice bath before being acidified to approximately pH 2.5 with aqueous HCl. The product filtered, which appeared as white precipitate was washed with distilled water and methanol before drying.

18.2.2.2   SYNTHESIS OF MOF MN

In order to form MOF structure, the metal linker ligands was attached to a metal linker such as manganese(II) acetate. Ideally the attachment will form a cubic structure. During this step, simple heating or post-synthetic modification was done to remove the Boc group from the amine. $Mn(Oac_3)_2.2H_2O$ (724 mg) in seven equal portions was added into the MOF ligand with a time delay of 15 minutes between each additions. After the last portion was added, the temperature was kept at 110 °C for 3 hours under stirring and additional 16 hours without stirring. After cooling down to room temperature, the solid was isolated by filtration, washed with 20 mL of DMF and ethanol and purified by soxhlet using 100 mL ethanol for 24 hours. After drying for 24 hours at 90 °C in an oven, the sample was stored in desiccators.

18.2.2.3   CATALYTIC EXPERIMENT

Figure 18.2 shows the catalytic performance scheme of Mn-MOF as oxidative catalysis. The aldehydes used in this study were 3-nitrobenzaldehyde and 2,4-hydroxybenzaldehyde.

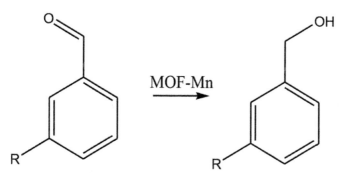

FIGURE 18.2   Oxidation reaction using $Mn(OAc)_3$-MOF as catalyst.

18.2.2.4   CHARACTERIZATIONS

The IR spectra of ligands and MOFs product were recorded using FTIR spectrophotometer with range of 4000–400 cm^{-1}. The sample was prepared using potassium bromide (KBr) pellet containing 1% finely ground samples.

18.3   RESULTS AND DISCUSSION

In this study, (Boc) was added to compound containing amine groups in the presence of DMAP. The Boc acted as protective groups for amine during the synthesis of ligand to prevent the interpenetration of structures. However, the Boc needed to be removed during the synthesis of framework or crystallization process. Thus, the addition of acid such as HCl was able to remove the Boc. The removal of Boc can increase the minimum fixed pore diameter of MOFs [5]. The attachment of manganese(II) acetate as metal linkers was used as catalyst for the oxidation reaction.

18.3.1   IR SPECTRA ANALYSIS

In order to determine the change that occurred in this study, FTIR analysis was chosen. The entire spectrum showed different patterns. Figure 18.3, (A) and (B), represents the spectrum of dimethyl 5-aminoisophthalate with Boc and aminoisopthalic respectively, while Figure 18.4 represents product formed after oxidation reaction. The medium of adsorption peaks at 1727.07cm^{-1}, Fig. 18.3(A) was assigned for υ(C=O). In the same spectrum, amine can be showed peak at 3369.1 cm^{-1} which assigned for aromatic secondary amine. Figure 18.3 (B) shows the different pattern of spectrum A especially at peak 3200 cm^{-1} that assigned for υ(OH). The peaks of spectrum in Figure 18.4 can be characterized as follows: both were successfully oxidized to alcohol group. Peak of spectra in Figure 18.4(i) was assigned for 3446.68 cm^{-1} for υ(OH), 1027.73 cm^{-1} for υ(C-OH), 1344.17 cm^{-1} for nitro group, while 1412.86 cm^{-1} and 1561.27 cm^{-1} were assigned for aromatic groups. In spectra (ii), peaks were assigned as υ(OH) at 3447 cm^{-1}, sharp and strong peaks at 1025.50 cm^{-1}

for υ(C-OH), while 1691.71 cm^{-1} and 1418.15 cm^{-1} both were assigned for aromatic compound.

TABLE 18.1 Wavenumber for Metal Linker Ligand: (A) Dimethyl 5-aminoisophthalate and (B) Aminiisophtalic Acid.

Figure 18.3	Wavenumber (cm^{-1})		
	υ(NH)	υ(C=O)	υ(OH)
A	3369.1	1727.07	–
B	3457.3, 3241.5	1641.22	3200

TABLE 18.2 Wavenumber for Product of Oxidation Aldehyde: (i) 3-Nitrophenol and (ii) 2,4- Dihydroxyphenol.

Figure 18.4	Wavenumber (cm^{-1})			
	υ(NO)	Aromatic Groups	υ(C-OH)	υ(OH)
i	1344.17	1561.27	1027.73	3446.68
ii	–	1691.71	1025.50	3447

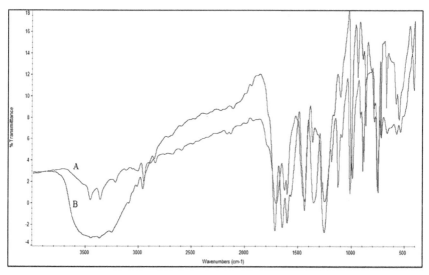

FIGURE 18.3 FTIR spectrum for metal linker ligand: (A) 5-aminoisophthalate and (B) aminoisopthalic acid.

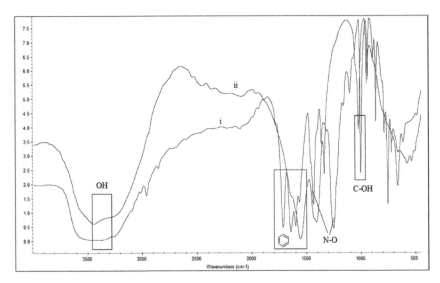

FIGURE 18.4 FTIR spectrum for product of oxidation of aldehyde (i) 3-nitrophenol and (ii) 2,4-dihydroxyphenol.

18.3.2 SCANNING ELECTRON MICROSCOPE (SEM)

Morphological structure of MOFs was revealed by scanning electron microscope (SEM) (Fig.18.5). SEM is capable of sample analysis as well as selected point location of sample. Generally SEM is use to generate high-resolution images of shapes of sample or objects (SEI) and to show spatial variations in chemical composition. However, it also widely used to identify the phases in crystalline structures.

Figure 18.45(a) shows the crumpled structure due to Boc protection which surrounding the MOFs. The micrograph of MOFs with manganese(II) acetate as a linker was observed in Figure 18.45(b). The cubic structure in Figure 18.54(b) showed the manganese acetate was successfully attached to a metal-organic linker ligand, D5P thus form the cubic structure. According to [9], bulky protective groups can preclude frameworks interpenetration as well as producing wide pore of open networks that suitable for application in storage and catalysis.

FIGURE 18.5   SEM illustration of MOF: (a) synthesized with Boc and (b) Mn-MOF.

18.4   CONCLUSION

This study was performed to synthesize and characterize the amino-based MOFs and to demonstrated oxidation reaction of manganese acetate. The using of Boc as protection for amine group was found to avoid steric clashes between NHBoc groups of neighboring ligands and increase the growth of MOFs. Characterization of MOFs ligands using FTIR Spectroscopy showed changes of absorption peaks in the spectrums. As oxidation reaction occurred, FTIR spectrum showed formation of carboxylic group. The cubic structure of the MOF as in shown of scanning electron microscope proved that MOF was successfully synthesized. For further research, the application of MOF should be tested in different carbon-carbon cross coupling reaction such as Suzuki Miyaura reaction complemented with leaching and recyclability tests.

18.5   ACKNOWLEDGMENT

The authors would like to express their gratitude to the Universiti Teknologi MARA for providing the facilities and DANA RIF [600-RMI/DANA 5/3/ RIF (759/2012)] and FRGS [600-RMI/FRGS 5/3 (52/2013)] for the financial supports.

KEYWORDS

- amino-based
- catalyst
- di-*tert*-butyl dicarbonate
- linker ligand
- metal-organic frameworks

REFERENCES

1. Li, J. R.; Kuppler, R. J.; Zhaou, H. C. *Chem. Soc. Rev.* **2009**, *38*, 1477–1504.
2. Hartman, M.; Fischer, M. Amino-functionalized Basic Catalyst with MIL-101 Structure. *Micropor. Mesopor. Mat.* **2012**, *164*, 38–43.
3. Kuppler, R. J.; Timmons, D. J.; et. al. Potential Application of Metal-Organic Framework. *Coor. Chem. Rev.* **2009**, *253*, 3042–3066.
4. Gason, J.; Aktay, U.; Hernendez-Alonso, M. D.; Gerard van Klink, P. M.; Kapteijin, F. Amino-Based Metal-Organic Framework as Stable Highly Active Basic Catalyst. *J. Catal.* **2009**, *261*, 75–87.
5. Despande, R. K.; Minnaar, J. L.; Telfer, S. G. *Angew. Chem. Int. Ed.* **2010**, *49*, 4598–4602.
6. Hamidipour, L.; Farzaneh, F. Cobalt Metal Organic Frameworks as an Efficient Heterogeneous Catalyst for the Oxidation of Alkanes. *React. Kinet. Cat.* **2013**, *109*, 67–75.
7. Ma, S.; Meng, L. Energy-Related Applications of Functional Porous Metal-Organic Frameworks. *Pure. Appl. Chem.* **2011**, *83*,167–188.
8. Song, F.; Zhang, T. C.; Wang, C.; Lin, W. Chiral Porous Metal-Organic Frameworks with Dual Active Sites for Sequential Asymmetric Catalysis. *Proc. R. Soc. A.* **2012**, *468*, 2035–2052.
9. Lun, D. J.; Waterhouse, G. I. N.; Telfer, G. S. A General Thermolabile Protecting Group Strategy for Organocatalytic Metal-Organic Frameworks. *J. Soc.* **2011**, *133*, 5806–5809.

CHAPTER 19

CONDUCTIVITY STUDIES OF SCHIFF BASE LIGANDS DERIVED FROM PHENYLENEDIAMINE DERIVATIVES

KARIMAH KASSIM[1,*] and MUHAMAD FARIDZ OSMAN[2]

[1]Institute of Science, Universiti Teknologi MARA, 40450 Shah Alam, Selangor, Malaysia

[2]Faculty of Applied Sciences, Universiti Teknologi MARA, 40450 Shah Alam, Selangor, Malaysia

*Corresponding author. E-mail: karimah@salam.uitm.edu.my

CONTENTS

ABSTRACT

Three Schiff base ligands, namely, 6,6'-(1E,1'E)-[1,2-phenylenebis(azan-1-yl-1-ylidene)]bis(methan-1-yl-1-ylidene)bis(2-methoxyphenol) **(L1)**, 6,6'-(1E,1'E)-[1,3-phenylenebis(azan-1-yl-1-ylidene)]bis(methan-1-yl-1-ylidene)bis(2-methoxyphenol) **(L2)**, and 6,6'-(1E,1'E)-[1,4-phenylenebis(azan-1-yl-1-ylidene)]bis(methan-1-yl-1-ylidene)bis(2-methoxyphenol) **(L3)** have been successfully synthesized using reflux condensation method. These ligands were analyzed using impedance measurements and optical properties for conductivity studies. Ligand L1 derived from *o*-phenylenediamine has higher conductivity value due to the presence of electron donating group and position of nitrogen donor atom in the structure.

19.1   INTRODUCTION

The chemistry of Schiff base ligands have attracted attention of many researchers due to their facile synthesis and wide range of applications such as antifungal, antibacterial, anticancer, anti-inflammatory agents, insecticidal, and catalytic field [1–3]. Other properties are spectral and thermal properties [4] as well as fluorescence properties to be applied in a wide range of new technologies [5]. These compounds are still found to be of great interest although they have been studied widely [6]. There are many studies on Schiff base compounds and metal complexes over the past decades. The availability of different types of amines and carbonyl compounds makes it possible to synthesize Schiff bases with variety structural features. Recent publications show few works on conductivity studies of cyclic conjugated Schiff base ligands. The synthesized compounds should have π conjugate bonds to become organic semiconductors which allow electrons move via π-electron cloud overlaps. Many researchers studied on acyclic and non-conjugated Schiff bases which lead to nonplanar conformation structure. Therefore, in this research, conjugated cyclic Schiff bases and their metal complexes have been synthesized to study the electrical conductivity of the compounds which are organic semiconductor compounds. In this chapter, we report the synthesis, characterization, and conductivity

studies   of   6,6'-(1E,1'E)-[1,2-phenylenebis(azan-1-yl-1-ylidene)]bis
(methan-1-yl-1-ylidene)bis(2-methoxyphenol)   (L1),   6,6'-(1E,1'E)-
[1,3-phenylenebis(azan-1-yl-1-ylidene)]bis(methan-1-yl-1-ylidene)
bis(2-methoxyphenol) (L2) and 6,6'-(1E,1'E)-[1,4-phenylenebis(azan-
1-yl-1-ylidene)]bis(methan-1-yl-1-ylidene)bis(2-methoxyphenol) (L3).
Conjugated compounds were chosen since they are more stable with
an effective conjugation system [7–9]. All ligands were synthesized
using the same aldehyde which is 2-hydroxy-3-methoxybenzaldehyde.
This study may improve semiconducting properties of Schiff bases and
enhance the application of conjugated Schiff bases in electronic devices.
The finding from this study provides a good basis for further application
in semiconductors industries.

19.2   MATERIALS AND METHODS

Schiff base ligands were prepared using 2-hydroxy-3-methoxybenzalde-
hyde, *o*-phenylenediamine, *m*-phenylenediamine, and *p*-phenylenediamine.
Solvents used for synthesizing and washing were ethanol and diethyl ether.
All the chemicals and solvents were purchased from commercial sources
and used as received without any further purification.

19.2.1   PREPARATION OF SCHIFF BASE LIGAND

A Schiff base ligand 6,6'-(1E,1'E)-[1,2-phenylenebis(azan-1-yl-1-ylidene)]
bis(methan-1-yl-1-ylidene)bis(2-methoxyphenol) (L1) was successfully
synthesized by condensation of 2-hydroxy-3-methoxybenzaldehyde and
o-phenylenediamine in 2:1 ratio. Both starting materials were dissolved in
25 cm^3 ethanol separately. The yellow solution of 2-hydroxy-3-methoxy-
benzaldehyde was added to *o*-phenylenediamine solution. The resulting
orange mixture was refluxed for 6 h. Then, the product was filtered off
and washed with diethyl ether. It was then left to dry for few days and
the dried product was collected. Same procedures were repeated by using
m-phenylenediamine and *p*-phenylenediamine for L2 and L3. Figure 19.1
shows the reaction scheme of L1 while Figure 19.2 shows the structures
of L2 and L3.

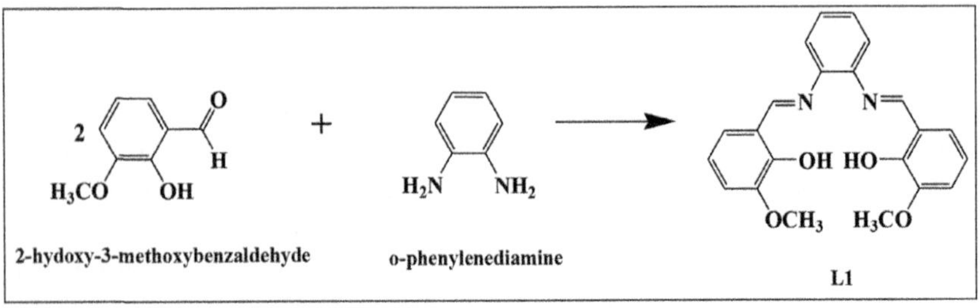

FIGURE 19.1 Reaction scheme of L1.

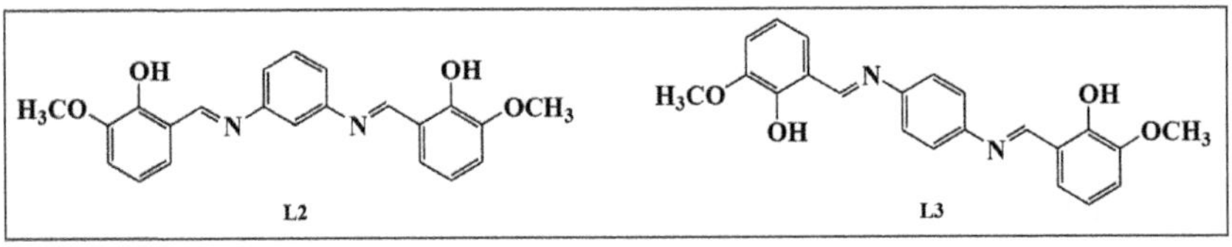

FIGURE 19.2 Structure of L2 and L3.

19.3 RESULTS AND DISCUSSION

19.3.1 ELEMENTAL ANALYSIS

The physical characteristics of the Schiff base ligands are shown in Table 19.1. For the elemental analysis, the experimental value was good agreement with the theoretical one. The melting point was in the range of 170–223°C. The ligands were produced in high yields.

TABLE 19.1 The Physical Characteristics of the Schiff Base Ligands.

Compound	Color	Melting Point (°C)	% Yield	Experimental Value (Calculated Value) %		
				C	H	N
L1	Orange	170.0	73.67	70.08	5.22	7.70
				(70.20)	(5.36)	(7.44)
L2	Orange	131.3	85.65	69.62	5.37	7.54
				(70.20)	(5.36)	(7.44)
L3	Orange	223.9	94.85	69.62	5.40	7.52
				(70.20)	(5.36)	(7.44)

19.3.2 INFRARED SPECTROSCOPY

Infrared spectroscopy was used to determine part of structural information about a molecule. It can be determined by the presence or absence of particular functional groups in the structure. This analysis was done on Perkin–Elmer 1750X FTIR. Table19.2 below summarizes the important peaks at different frequency that appears in the spectra. In the IR spectrum of ligand L1, a peak at 1614 cm^{-1} attributed to the C=N stretching band in the structure while for L2 and L3 are 1618 cm^{-1} and 1607 cm^{-1}, respectively. The presence of C=N stretching frequencies proved the formation of imine group. The phenolic C–O stretching appeared at 1252–1256 cm^{-1} for all ligands. Table 19.3 summarizes all the IR bands in the spectra of all compounds.

TABLE 19.2   The Stretching Frequencies of Schiff Base Ligands.

Compound	Wavelength (cm−1)	
	$v(C = N)$	$v(C–O)$
L1	1614	1256
L2	1618	1252
L3	1607	1255

19.3.3 ¹H NMR SPECTROSCOPY

The ^{1}H NMR spectra were recorded in CDCl$_3$ solution using TMS as internal standard. L1, L2, and L3 showed a singlet signal between 8.61 ppm and 8.70 ppm which can be assigned to azomethine (N=CH) proton. The signal of phenolic proton was observed as a singlet at the range of 13.20–13.60 ppm. A singlet sharp signal indicated methoxy proton appeared in the spectra at 3.89 ppm, 3.94 ppm, and 3.90 ppm for L1, L2, and L3, respectively. Multiple signals were observed at 6.83–7.34 ppm for L1, 6.87–7.49 ppm for L2, and 6.90–7.40 ppm for L3 due to aromatic protons.

TABLE 19.3 ^{1}H NMR Spectroscopic Data.

Compound	Chemical Shift, δ (ppm)			
	N=CH	O–H	OCH$_3$	Aromatic
L1	8.61	13.20	3.89	6.83–7.34
L2	8.67	13.47	3.94	6.87–7.49
L3	8.70	13.60	3.90	6.90–7.40

19.3.4 CONDUCTIVITY MEASUREMENT

Conductivity of all compounds was measured using impedance analyzer by applying alternating current (ac) voltage. The samples (0.25 g) were prepared in the form of pellets and the thickness was measured. The pellets were placed between two stainless steel blocking electrodes. The ac impedance spectroscopy (IS) was measured by using Solartron Impedance 1260 in the frequency range of 0.1 Hz to 32.7 MHz at room temperature. Conductivity value was calculated using this formula:

$$\frac{T}{R_b A}, \tag{19.1}$$

where

T = thickness of the pellet,
R_b = bulk resistance,
A = surface area of the pellet.

TABLE 19.4 Conductivity Value

Compound	Conductivity (S/cm)
L1	1.37×10^{-7}
L2	6.53×10^{-8}
L3	3.48×10^{-8}

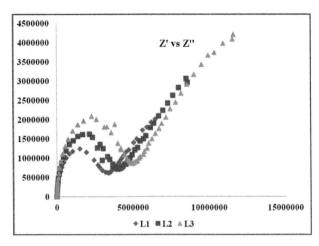

FIGURE 19.3 Impedance spectra of L1, L2, and L3.

19.3.5 OPTICAL PROPERTIES

The optical property represents the band gap energy was done to support the conductivity data. The result can be acquired by conducting UV-Vis experiment. Optical band gap values can be obtained by using this equation:

$$\frac{hC}{\lambda_{onset}(1.6 \times 10^{-19})},$$ (19.2)

where

λ_{onset} = wavelength value from two tangents on the absorption edges (it also indicates the electronic transition start wavelength),

h = Planck's constant,

C = speed of light.

The optical band gap energy is the energy taken by an electron to move from HOMO to LUMO. Figure 19.4 shows determination of λ_{onset} on L1. From the UV/vis spectrums, the optical band gap energy supports the conductivity values as the large optical band gap indicates more difficult electron to excite from HOMO to LUMO. Hence, it reduces the conductivity value. Conductivity value increases with decreasing band gap energy. Table 19.5 shows the band gap energy of all synthesized compounds.

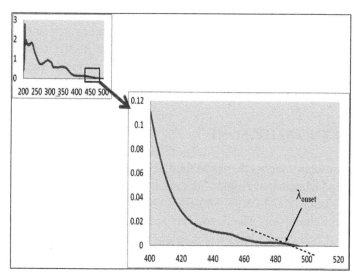

FIGURE 19.4 Absorption spectrum and λ_{onset} of L1.

TABLE 19.5 Band Gap Energy for Schiff Base Ligands.

Compound	Band Gap Energy (eV)
L1	1.56
L2	2.56
L3	2.76

Besides the explanation on HOMO and LUMO, conjugation of structure of the compounds also affect the conductivity values as well as the presence of electron withdrawing and electron donating groups. According to Orton (2009) [10], organic semiconductor compounds that contain conjugated bands allow the movement of electrons in the structure. However, the presence of electron withdrawing and electron donating groups alter the distribution on the conjugated structure of organic semiconductors as the increasing or decreasing of electron density to the system by these substituent groups [11]. In 1965, Gooden [12] stated that the presence of unpaired electron at azomethine nitrogen (C=N) affected the electrical conductivity values.

19.4 CONCLUSION

Three Schiff base ligands were successfully synthesized and characterized. Conductivity studies showed that L1 has higher conductivity value followed by L2 and L3. Band gap energy supports the conductivity data.

19.5 ACKNOWLEDGMENT

The author expresses his thanks to the Faculty of Applied Sciences for the funding and Ministry of Higher Education (MOHE) for the research grant FRGS/1/10/SG/UITM/03/05, 600-RMI/DANA 5/3/PSI (37/2013) and for MyMaster Scholarship. Thank you also to Faculty of Applied Sciences and Institute of Science Universiti Teknologi MARA for providing research facilities.

KEYWORDS

- conductivity
- ligands
- Schiff base
- phenylenediamine
- band gap energy

REFERENCES

1. Arish, D.; Nair, M. S. Synthesis, Spectroscopic, Antimicrobial, DNA Binding and Cleavage Studies of Some Metal Complexes Involving Symmetrical Bidentate N, N Donor Schiff Base Ligand. *Spectrochim. Acta. A* **2011**, *82*, 191–199.

2. Patil, S.; Jadhav, S. D.; Patil, U. P. Natural Acid Catalyzed Synthesis of Schiff Base Under Solvent-Free Condition: As a Green Approach. *Arch. Appl. Sci. Res.* **2012**, *4*(2), 1074–1078.

3. Al-Ali, K. A. H. Synthesis, Characterization and Study of Electrical Properties of Fe(III) and Al(III) Complexes of a Schiff Base. *J. Univ. Thi-Qar* **2011**, *6*(3), 1–19.

4. Sallam S. A.; Ayad M. I. *J. Korean Chem. Soc.* **2003**, *47*(3), 199–205.

5. Guo, L.; Wu, S.; Zeng, F.; Zhao, J. Synthesis and Fluorescence Property of Terbium Complex with Novel Schiff-base Macromolecular Ligand. *Eur. Polym. J.* **2006**, 1670–1675.

6. Thakor, Y. J.; Patel, S. G.; Patel, K. N. Synthesis, Characterization and Biocidal Studies of Some Transition Metal Complexes Containing Tetradentate and Neutral Bidentate Schiff Base. *J. Chem. Pharm. Res.* **2010**, *2*(5), 518–525.

7. Abbas, S. A.; Munir, M.; Fatima, A.; Naheed, S.; Ilyas, Z. Synthesis and Analytical Studies of Sulfadimidine-imine Schiff Base Complexes with Ni(II) and Cu(II). *J. Life Sci.* **2010**, 37–40, ISSN: 2078-5291.

8. Chandra, S.; Sangeetika, X. EPR, Magnetic and Spectral Studies of Copper(II) and Nickel(II) Complexes of Schiff Base Macrocyclic Ligand Derived from Thiosemicarbazide and Glyoxal. *Spectrochim. Acta. A Mol. Biomol. Spectrosc.* **2004**, *60*(1–2), 147–153.

9. Mederos, A.; Dominguez, S.; Hernandez-Molina, R.; Sanchiz, J.; Brito, F. Coordinating Ability of Ligands Derived from Phenylenediamines. *Coordin. Chem. Rev.* **1999**, *193–195*, 857–911.

10. Orton, J. *Semiconductors and the Information Revolution;* Academic Press: Amsterdam, 2009, pp. 18–27.

11. Germain, J. P.; Pauly, A.; Maleysson, C.; Blanc, J. P.; Schollhorn, B. Infuence of Peripheral Electron-Withdrawing Substituents on the Conductivity of Zinc Phthalocyanine in the Presence of Gases. Part 2: Oxidizing Gases. *Thin Solid Films* **1998,** *333,* 235–239.

12. Gooden, E. Organic Semiconductors: Effect of Substituents and Metal Ions on Electrical Conductivity of Schiff Base. *Aust. J. Chem.* **1965,** *18,* 637–650.

INDEX

Printed and bound by CPI Group (UK) Ltd, Croydon, CR0 4YY

23/10/2024

01777705-0002